PIEZOELECTRICITY - ORGANIC AND INORGANIC MATERIALS AND APPLICATIONS

Edited by **Savvas G. Vassiliadis** and **Dimitroula Matsouka**

Piezoelectricity - Organic and Inorganic Materials and Applications
http://dx.doi.org/10.5772/intechopen.72336
Edited by Savvas G. Vassiliadis and Dimitroula Matsouka

Contributors

Huacheng Qiu, Helmut Seidel, Sampo Tuukkanen, Satu Rajala, Hirokazu Madokoro, Hiroki Takise, Masato Suzuki, Tomokazu Takahashi, Seiji Aoyagi, Seungbum Hong, Melodie Glasser, Panpan Li, Jeongjae Ryu, Dimitroula Matsouka, Savvas G. Vassiliadis, Boris Sorokin, Gennadiy Kvashnin, Andrey Novoselov, Sergey Burkov, Anton Shipilov, Nikolay Luparev, Victor Aksenenkov, Vladimir Blank

First published in London, United Kingdom, 2018 by IntechOpen
IntechOpen is the global imprint of INTECHOPEN LIMITED, registered in England and Wales, registration number: 11086078, The Shard, 25th floor, 32 London Bridge Street
London, SE19SG – United Kingdom
Printed in Croatia

British Library Cataloguing-in-Publication Data
A catalogue record for this book is available from the British Library

Additional hard copies can be obtained from orders@intechopen.com

Piezoelectricity - Organic and Inorganic Materials and Applications, Edited by Savvas G. Vassiliadis and Dimitroula Matsouka
p. cm.
Print ISBN 978-1-78923-646-0
Online ISBN 978-1-78923-647-7

For EU product safety concerns:
IN TECH d.o.o., Prolaz Marije Krucifikse Kozulić 3, 51000 Rijeka, Croatia,
info@intechopen.com or visit our website at intechopen.com.

Meet the editors

Prof. Savvas Vassiliadis holds degrees in Textiles, Electronics, and Electrical and Computer Engineering and his PhD degree in Computational Modeling. After a long industrial career, he joined the Department of Electrical and Electronics Engineering of the University of West Attica, Athens, Greece, where he served as the head of the department for 4 years, a member of the Administration Council, and the vice rector for International Relations. He is the author of more than 150 scientific articles published in international journals and presented at international conferences. He has published 7 books and participated in more than 35 national and international research and technological projects. He is a member of the scientific committees of many international conferences and editorial boards of journals.

Dr. Dimitroula Matsouka holds degrees in Textile Engineering and Quality Assurance and her PhD degree in Materials Science. She has more than 15 years' experience in the field of quality control and research in Textiles, including analysis of technical problems encountered in production situations. Her publication history includes more than 20 articles in national and international journals, international conferences, and book chapters. She has participated in 11 national and international research and technological projects. She is a member of the Marie Curie Alumni Association and the Balkan Society of Textile Engineers.

Contents

Preface

Piezoelectricity, from the Greek *piezein*, meaning to press (squeeze), and *ilektron*, meaning amber, a material with electrostatic properties that were observed since ancient times, is a process in which a material generates electric voltage in response to mechanical deformation. Theophrastus (fourth century BC), in his treatise "On Stones," observed that a mineral, probably modern-day tourmaline, attracted straws and bits of wood. Subsequently, in 1880, French scientists the Curie brothers described their qualitative observations of the piezoelectric behavior of certain crystals—mainly not only quartz but also tourmaline—in a letter to Comptes Rendus de I'Academie des Sciences, the proceedings of the French Academy of Science. The Curie brothers are credited with the discovery of piezoelectricity in the modern age. The first commercial application of piezoelectricity came with WWI when Paul Langevin and his coworkers (1917) developed an ultrasonic submarine detector (sonar).

Historically, the timeline of piezoelectric materials follows the order, crystals (quartz), then piezoceramics ($BaTiO_3$), and later piezoelectric polymers (PVDF). The history of piezoelectric materials and applications also reads in parallel with the history of WWI and WWII, from sonar to the requirement for a replacement for mica in capacitors due to fears of U-boats threatening the supply from South America to the United States.

From those origins, piezoelectricity has expanded into a great variety of applications and a great variety of materials. In inorganic materials, $BaTiO_3$ has largely been replaced by PZT (the most widely used and researched piezoceramic material), and there have been attempts at replacing PZT with lead-free materials, namely perovskite [i.e., $BiNaTiO_3$ (BNT), $KNaNbO_3$ (KNN), etc.], non-perovskite bismuth layer-structured ferroelectrics (BLSF), and non-perovskite tungsten-bronze-type ferroelectrics.

In organic materials, the most commonly used and widely researched PVDF is being supplemented by its copolymer with trifluoroethylene (TrFE), carbon black, or carbon nanotubes. These additives are used to enhance the crystal orientation. Copolymers of PVDF with inorganic piezoelectric materials, such as $BaTiO_3$ and PZT, also exist combining the properties of both types of materials. Voided charged polymers, such as cellular polypropylene where the voids can be filled either with ambient air or with selected gases like nitrogen, are used in applications such as microphones.

This great variety of materials being researched, developed, or already in use can only mirror the great number of fields, where piezoelectric materials are used. Roughly, there are three main areas: sensors, actuators, and energy-harvesting applications. An interesting example for sensors is the tactile sensor applications. Tactile sensors (sensors that measure through contact or touch) can be used to measure temperature, shape, softness, pressure, or force. The case of tactile sensing is significant because the use of polymers over inorganic materials is a one-way street due to the mechanical flexibility and softness needed.

Piezoelectric actuators can work in tandem with piezoelectric sensors and/or separately. For example, piezoelectric actuators can provide tactile sensations that can be used either as a means of feedback (e.g., tactile perceptions from a glove in a virtual reality application) or as information (e.g., indicating the direction of safety and least temperature inside a firefighter's uniform by vibrating against the firefighter's arms). Another example is the use of piezoelectric actuators in robotics and mechatronics. These actuators are used as positioners, motors, and vibration dampers in applications such as optical devices, small-scale ultrasonic motors, and active vibration controllers in space structures.

Energy-harvesting applications are concerned with the harvesting of "waste" energy from the environment, and in the case of piezoelectric materials, energy produced through vibrations is the most common. A popular area regarding the possibility of harvesting energy as a byproduct of some activity is the research into harvesting energy from the movement of the human body, especially walking—with shoe inserts in the heel area being the most researched approach. This field fits quite well with the interest in wearable smart materials and applications in a two-in-one strategy. A significant problem that has been relatively neglected up to now regarding vibration energy harvesters is the matching between the harvester input impedance and the output load impedance. Usually, microelectromechanical system (MEMS) designers of energy harvesters report the power at the optimal output load or the output open-circuit voltage as output power.

This book "Piezoelectricity: Organic and Inorganic Materials and Applications" consists of seven high-level chapters embracing the major application fields of the piezoelectric materials.

The first chapter "Nanocellulose as a Piezoelectric Material" examines the piezoelectric properties of nanocellulose, a biobased nanomaterial. Nanocellulose is studied in the form of film with very promising experimental results.

The second chapter "Application of Thin Piezoelectric Films in Diamond-Based Acoustoelectronic Devices" examines the behavior of thin piezoelectric films, especially in the GHz frequency range. This chapter successfully combines theory with experimental results.

The third chapter "Hydrodynamic Loading on Vibrating Piezoelectric Micro Resonators" discusses microresonators vibrating in fluid. This chapter investigates the different resonant modes as well as the influences of fluid viscosity and compressibility in free space and close to a surface.

The fourth chapter "Piezoelectric Melt-Spun Textile Fibers: Technological Overview" is an in-depth technological review of the area of melt-spun piezoelectric textile fibers. The piezoelectric textile fibers are of high interest because of their potential use in wearable electronics.

The fifth chapter "Piezoelectric Vibration Energy Harvester Using Polyvinilidene Difluoride Film Formed by Bar-Coating Method and Its Spray-Coating Method on Three-Dimensional Surface" examines the fabrication process of a vibration energy harvester and presents the related experimental results.

The sixth chapter "Piezoelectric Sensors Used for Daily Life Monitoring" presents a system for the monitoring of daily human activities using sensors, especially piezoelectric ones. This chapter evaluates the prototype system and presents promising results.

The seventh chapter "Piezoelectric Materials for Medical Applications" is an extended review of the behavior of organic piezoelectric tissues of the human body. It also examines the use of piezoelectric materials in medical implants and equipment.

These seven chapters present modern aspects and advances in the field of piezoelectricity, and the collective publication of them provides an integrated study for inspiring scientists in the field. This book is an influential contribution in piezoelectricity.

Prof. Dr. Savvas Vassiliadis
University of West Attica
Athens, Greece

Dr. Dimitroula Matsouka
University of Bolton
Bolton, UK

Chapter 1

Nanocellulose as a Piezoelectric Material

Sampo Tuukkanen and Satu Rajala

Additional information is available at the end of the chapter

http://dx.doi.org/10.5772/intechopen.77025

Abstract

Cellulose-based nanomaterials, which are generally known as nanocelluloses, are interesting renewable biomaterials which have potential applications for example in material science, electronics and biomedical engineering and diagnostics. Cellulose has a strong ability to form lightweight, highly porous and entangled networks that make nanocellulose suitable as substrate or membrane material. Recently, also studies related to piezoelectric behavior of nanocellulose have been published. The piezoelectricity of wood was proposed already in 1955 by Eiichi Fukada, but only very slightly studied since then. Here, we show the experimental evidence of significant piezoelectric activity of different types of nanocellulose films. Wood-based cellulose nanofibril (CNF) and cellulose nanocrystals (CNC) films, and bacterial nanocellulose (BC) films have been studied. The recent results suggest that nanocellulose is a potential bio-based piezoelectric sensor material.

Keywords: nanocellulose, bacterial cellulose, cellulose nanofibril, piezoelectric sensor, cellulose nanocrystal

1. Introduction

Cellulose, the most abundant biopolymer on earth, is generally obtained from plant sources. Cellulose-based nanomaterials, such as wood-based cellulosic nanofibrils (CNF) and cellulose nanocrystals (CNC) as well as bacterial cellulose (BC), are also known as nanocelluloses [1]. These interesting bio-based and renewable nanomaterials have potential applications in many different fields. Their nanoscale dimensions as well as a strong ability to form entangled porous networks make nanocelluloses suitable source materials for lightweight membranes, films and nanopapers. Importantly, these materials can be processed in aqueous media, which ensures low cost and high throughput manufacturing of functional devices for electronics [2–4], sensing [5], and optics [6]. Further, combining of nanocelluloses with water-processable carbon nanomaterials, such as carbon nanotubes or graphene, open routes to fabricate flexible and disposable electronic devices, such as supercapacitors [7–9].

Figure 1 describes fabrication of CNF and CNC from wood cellulose fibers. In principal, wood-based nanocellulose is fabricated by cutting the wood cellulose fibers into smaller units until its dimensions become into nanoscale, referring to 'top-down approach'. CNF can be obtained, for example, from birch wood by sulfate digestion and subsequent bleaching, followed by miniaturization processing through a Masuko grinder and homogenization by microfluidizer [10].

On the other hand, BC films are produced through oxidative fermentation of polysaccharides by certain bacteria such as *Komagataeibacter xylinus,* which can be considered as 'bottom-up approach.' In comparison with wood-based cellulose, bacterial cellulose (BC) holds interesting properties such as biodegradability, high purity, water-holding capacity, and superior mechanical and structural properties. The free-standing films can be obtained from wood-based nanocellulose (CNF and CNC) by solution casting methods [11], while the BC films are produced as such as free-standing films.

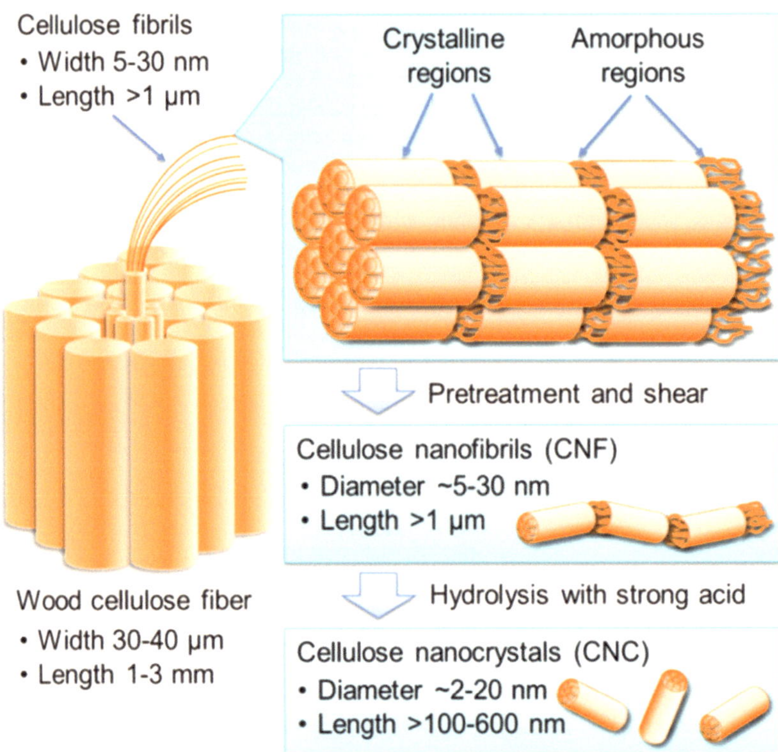

Figure 1. A schematic view of fabrication of nanocelluloses from wood cellulose fibers. The cellulose fibers are first cut up into microfibrils and finally after several processing steps into cellulose nanofibrils (CNF). Cellulose nanocrystals (CNCs) can be then obtained from CNF through acid hydrolysis. Nanocelluloses processing and structures are described in more detail in the literature [1].

In this chapter, we present the basics of nanocellulose as a piezoelectric material. Both the wood-based nanocellulose and bacterial nanocellulose films have been studied, and their sensitivities have been measured with a sophisticated sensitivity measurement setup. Finally, the results are presented, and the possibilities of nanocellulose as piezoelectric material are discussed.

2. Piezoelectricity of nanocellulose

The piezoelectricity of wood, that is, the change of electrical polarization in a material in response to mechanical stress, has been known for decades [12, 13]. The piezoelectric effect is highly enhanced if one considers the isolated crystalline building blocks of wood, CNCs [14, 15]. However, related issues have been covered in the scientific literature to a very limited extent, and only few recent reports discuss the experimental evidence of CNC piezoelectricity [14, 16].

The piezoelectricity is a phenomenon, where pressing of a piezoelectric film by an external force causes a change in charge distribution in the film. This change in charge distribution generates a measurable voltage between two electrodes placed on the top and the bottom of the piezoelectric film. The piezoelectric coefficient d_{mn} describes the charge density generated under a certain applied stress. The components of third-rank tensor with piezoelectric coefficients d_{mn} can be expressed using a 3 × 6 matrix, where $m = 1, 2, 3$ refers to the electrical axis and $n = 1, 2, \ldots, 6$ to the mechanical axis [17]. The main axes 1, 2 and 3 correspond to length, width and thickness, whereas the shear around these axes is expressed by indexes 4, 5 and 6.

In the case of a single cellulose crystal, piezoelectric tensor d_{mn} can be derived from the symmetry of a cellulose crystal lattice, formed by unit cells of cellulose molecules ($[C_6H_{10}O_5]_n$) [13]. The cellulose possesses a monoclinic symmetry with space group of $C_2 \parallel x_3$ having a following piezoelectric tensor:

$$d_{mn} = \begin{pmatrix} 0 & 0 & 0 & d_{14} & d_{15} & 0 \\ 0 & 0 & 0 & d_{24} & d_{25} & 0 \\ d_{31} & d_{32} & d_{33} & 0 & 0 & d_{36} \end{pmatrix}. \quad (1)$$

It is important to notice that this tensor is valid only for a single cellulose crystal. However, for an assembly of randomly aligned crystals, such as in a CNF film, the overall piezoelectric response results from the combination of the different coefficients. For example, in the case of wood, the arrangement of fibers and accompanied cellulose crystals have been shown to exhibit significantly reduced effective piezoelectric tensor:

$$d_{mn} = \begin{pmatrix} 0 & 0 & 0 & d_{14} & 0 & 0 \\ 0 & 0 & 0 & 0 & d_{25} & 0 \\ 0 & 0 & 0 & 0 & 0 & 0 \end{pmatrix}, \quad (2)$$

where $d_{14} = -d_{25}$ [12, 13].

3. Piezoelectric nanocellulose sensors

3.1. Nanocellulose film characterization

The nanocellulose films can be characterized with versatile methods. In our previous approaches, the nanocellulose films are characterized, for instance, by scanning electron microscopy and photometric stereoimaging to obtain information on the structure and surface topography of the films. In addition, dielectric properties such as relative permittivity and loss tangent, and ferroelectric hysteresis have been measured for CNF film [18]. Tensile stress tests have also been performed as well as crystallinity index measured for BC film [19].

The fabricated CNF films had a thickness of 45 ± 3 μm and density of 1.38 g/cm^3 [18]. For the BC films, the thickness of films varied from 4.8 ± 0.8 to 10.3 ± 0.6 μm depending on the film type. Different BC film types were produced in order to improve BC production titer and tailored alterations into the BC films [19]. To do this, we engineered *K. xylinus* to overexpress partial and complete BC synthase operon that encodes BC production activity. The changes in cell growth, end metabolite, and BC production titers obtained from the engineered strains (named pA, pAB and pABCD) were compared with the wild-type *K. xylinus* (named WT).

Figure 2 shows examples of SEM cross-section and surface view images of (a-b) CNF and (c-d) BC films. For the CNF film, the cross-section image reveals a layered a porous structure similar to reported elsewhere [11]. The surface view of the film (**Figure 2b**) shows a random

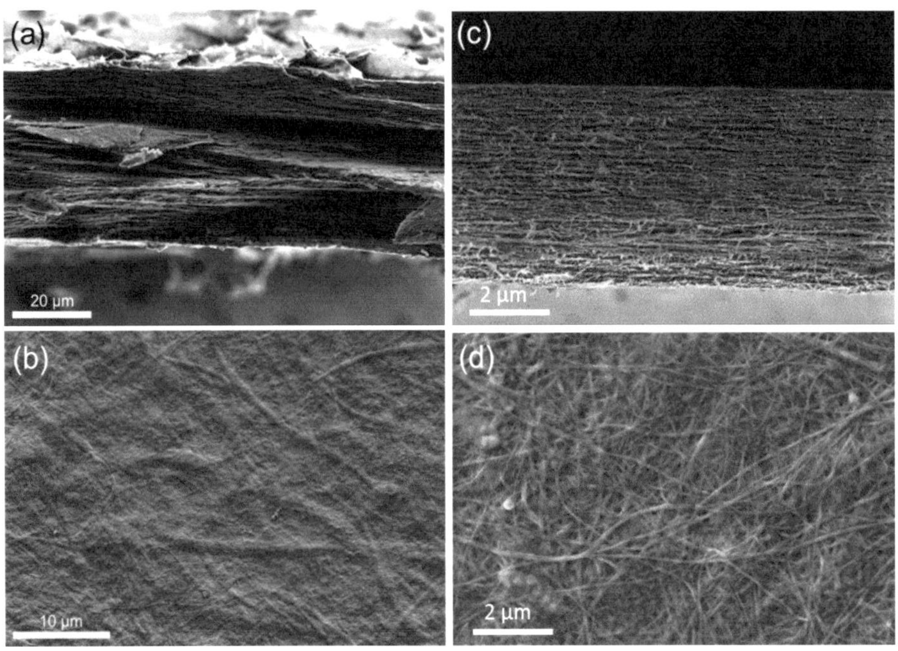

Figure 2. Scanning electron microscope cross-section and surface view images of typical (a, b) CNF [18] and (c, d) BC [19] films.

orientation of the nanofibrils. For the BC film, also a layered structure was found; however, the cross-sectional view of the films reveals that a layered fiber was more distinguished on the top surface of the films when compared with the bottom surface structure.

As the piezoelectricity is closely related to ferroelectricity of a material, the CNF films have been also analyzed by using a ferroelectric tester to measure the ferroelectric hysteresis of the film [18]. The results of ferroelectric hysteresis measurements are shown in **Figure 3**. The capacitance (or linear component in the polarization data) is dominant in the measurements at electric fields from 5 to 15 V/μm (**Figure 3a**). This shows that the CNF film has no significant ferroelectric hysteresis at low or moderate electric fields. However, a nonlinear behavior in polarization is detected at electric fields between 40 and 50 V/μm (**Figure 3b**). This suggests that the CNF film has a certain level of ferroelectric properties at high electric fields. A remanent polarization at 50 V/μm field was as small as 0.15 μC/cm^2.

The ferroelectric hysteresis measurements showed that the CNF film can withstand relatively high electric fields, despite its porous microstructure [18]. Small remanent polarization can be observed with electric fields above 40 V/μm which indicates that nanocellulose acts like an incipient organic ferroelectric material under a very high electric field [20]. The high coercive field in comparison with similar inorganic materials is a commonly known property of organic ferroelectric materials [20]. It should be remarked that at high electric fields, the electrostriction possibly resulting from the using of sandwiched measurements electrodes may also contribute to the polarization, while ferroelectricity is expected to be the dominant effect.

Importantly, our results show that the CNF films investigated in this work need a high electric field for poling and exhibit characteristics of ferroelectricity only above 40 V/μm. In that case, the observed ferroelectric (or piezoelectric) effect of the CNF films is expected to result from the permanent dipole moment of CNCs which possess a certain level of (nonintentional) orientation of dipoles resulting from the film fabrication process. However, the orientation of CNCs is not apparent from SEM images (**Figure 2**) due to limited resolution. There are recent related observations where experimental evidence of giant permanent electric-dipole moment in CNCs is reported [16].

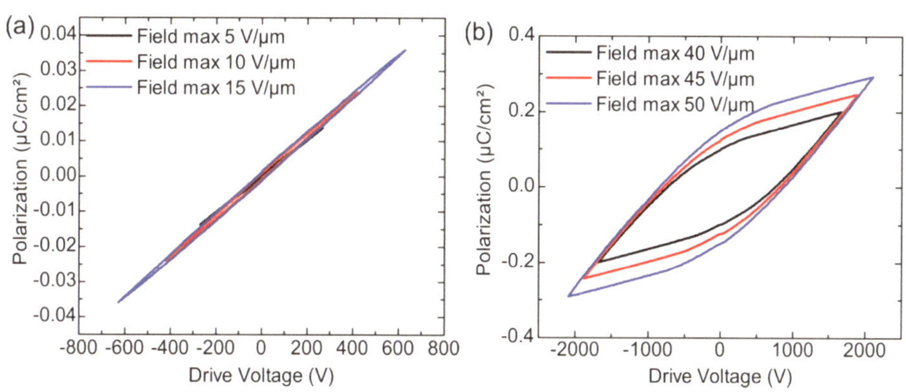

Figure 3. Polarization-voltage hysteresis curves for the CNF film at (a) 5–15 V/μm and (b) 40–50 V/μm electric fields at room temperature.

In addition, the relative permittivity and dielectric loss tangent for the CNF film have been determined [18]. At a 1 kHz frequency, the obtained relative permittivity of the CNF film was about 3.47, whereas the dielectric loss (tan δ) was about 0.011. The capacitance of the CNF film varied from 78 to 73 pF at frequencies from 100 Hz to 1 MHz, respectively. The average values for relative permittivity and dielectric losses were 3.38 and 0.071, respectively.

3.2. Sensor assembly

The sensors are typically assembled by sandwiching the nanocellulose film between two electrodes. In our approaches, the electrodes for the sensors were fabricated on 125-μm-thick polyethylene terephthalate (PET) substrate by e-beam evaporation. Typically, 100 nm thick layer of copper was evaporated through a shadow mask to provide electrode pattern. The nanocellulose film was then sandwiched between two electrodes, as illustrated in **Figure 4**. The connection to the measurement electronics was provided with crimp connectors (Nicomatic Crimpflex).

3.3. Piezoelectric sensitivity

The piezoelectric sensitivity measurement setup presented here has been previously used to evaluate piezoelectric response of a piezoelectric polymer film (polyvinylidene fluoride, PVDF)-based plantar pressure sensor [21, 22], PVDF sensors with solution-processed electrodes [23, 24],

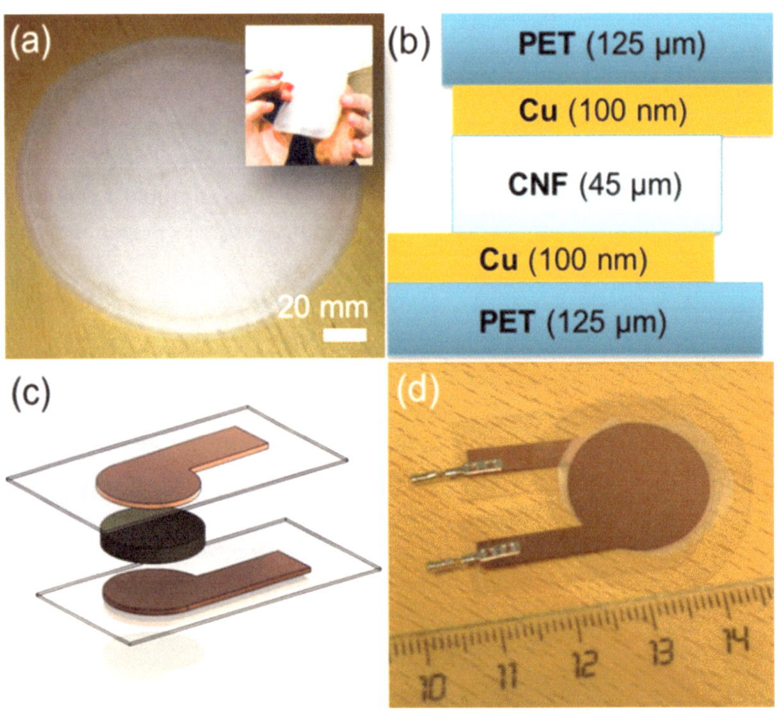

Figure 4. (a) Photographs of a fabricated self-standing CNF film and its bending robustness (an insert). (b, c) Schematic side view and (d) a photograph of assembled sensor.

http://dx.doi.org/10.5772/intechopen.77025

ferroelectret film-based sensors [25] and nanocellulose sensors [18, 19, 26–28]. This section briefly describes the setup.

The Brüel & Kjaer Mini-Shaker (Type 4810) used in the setup generates a dynamic excitation force. A sinusoidal input signal for the shaker was provided from a function generator (Tektronix AFG3101). A high sensitive dynamic force sensor (PCB Piezotronics, model number 209C02) was used as a reference sensor to measure the dynamic excitation force applied to the piezoelectric sensor sample under test. The dynamic force sensor was connected to a sensor signal conditioner (PCB Piezotronics, Model 442B06) with a low-noise coaxial cable. A load cell (Measurement Specialties Inc., model number ELFS-T3E-20 L) was used as a reference sensor to measure the static force applied to the sample by the shaker's piston. A pretension, which is producing static force, is required to hold the sample in place and to prevent the piston jumping off from contact during the measurement. A static force of approximately 3 N is typically used in the measurement setup.

The sensor sensitivity measured with the setup is closely related to the longitudinal piezoelectric coefficient d_{33}. The longitudinal d_{33} coefficient describes the electric polarization generated in the same direction as the stress is applied [29]. Thus, to measure the sensor sensitivity in normal force direction, the sensor was placed horizontally on the metal plate. The charge developed by the sensor was measured with a custom-made combination of a charge amplifier and a 16-bit AD-converter. **Figure 5a** illustrates the sensor sensitivity measurement setup.

A dynamic force of approximately 1.3 N (peak-to-peak), measured with the reference dynamic force sensor, was typically applied on the sample in the sensitivity measurements. This dynamic force was obtained by applying a dynamic, sinusoidal 2 Hz input signal of 1000 mV (peak-to-peak) to the shaker. To increase statistics, the sensitivity was measured from nine different positions on the sample, as shown in **Figure 5b**. The same positions were excited from both sides of the sample, resulting in a total of 18 excitations per sensor. Finally, the sensitivity was obtained by dividing the charge generated by the sensor with the force obtained with the dynamic force sensor. The unit of sensitivity is thus pC/N.

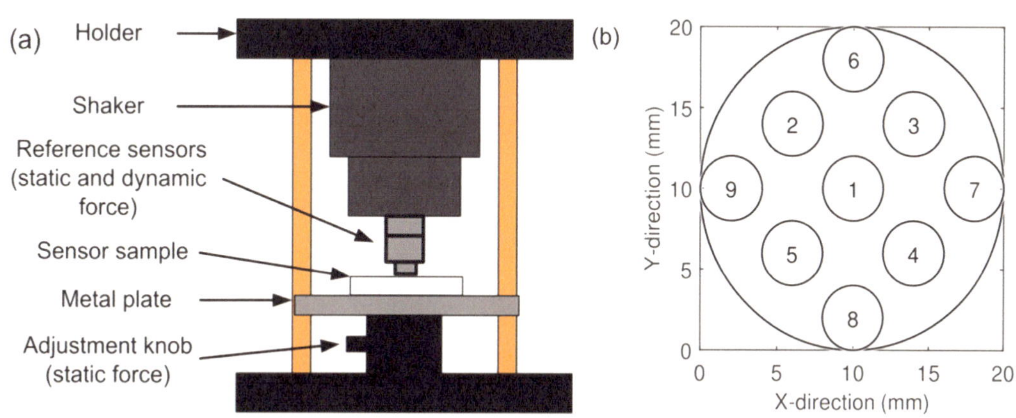

Figure 5. (a) Piezoelectric sensitivity measurement setup and (b) excitation positions for the sensitivity distribution measurements.

The same setup can also be used to measure the nonlinearity and hysteresis of the sensors. The sensor nonlinearity is determined by increasing the amplitude of the dynamic excitation force from approximately 0.1–5 N (peak-to-peak, frequency 2 Hz) and measuring the charge generated by the sensor. The sensor hysteresis error is defined as the deviation of the sensor output charge at a specified point of the excitation force when this point is approached from opposite directions (increasing or decreasing dynamic force) [30]. The excitation force range used was approximately from 0.7 to 2.8 N (peak-to-peak, frequency 2 Hz).

Table 1 summarizes the results of piezoelectric sensitivity measurements, shown as average values, obtained for the nanocellulose sensors and compared with reference PVDF sensors. The nonlinearity is defined as the maximum difference between the measured values and the fitted first degree polynomial, and the hysteresis as the maximum difference between the values measured when the excitation force was increased and decreased. The nonlinearity and hysteresis for CNF and reference PVDF sensors are further illustrated in **Figure 6**. **Figure 7** illustrates the BC and reference PVDF sensor charges as a function of increasing applied force, from which the nonlinearity presented in **Table 1** for BC can be obtained.

	Sensitivity (pC/N)	Nonlinearity (pC/N)	Hysteresis error (pC/N)
CNF [18]	4.7 ± 0.9	0.9 ± 0.5	0.9
CNF [27]	2.2 ± 1.5	—	—
CNC [27]	7.3 ± 2.5	—	—
BC WT [19]	16.8 ± 3.9	0.4 ± 0.1	—
BC pA [19]	7.3 ± 1.6	—	—
BC pAB [19]	5.0 ± 1.9	—	—
BC pABCD [19]	8.2 ± 5.5	—	—
PVDF [18]	27.5 ± 2.6	6.5 ± 3.8	0.9

Table 1. Summary of results for the sensor sensitivity, nonlinearity and hysteresis error measurements [18, 19, 27].

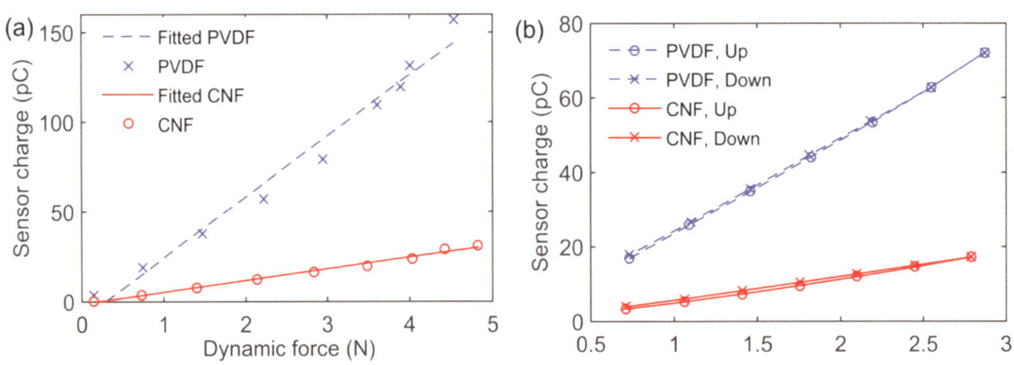

Figure 6. (a) Nonlinearity and (b) hysteresis curves measured for the CNF and reference PVDF sensors.

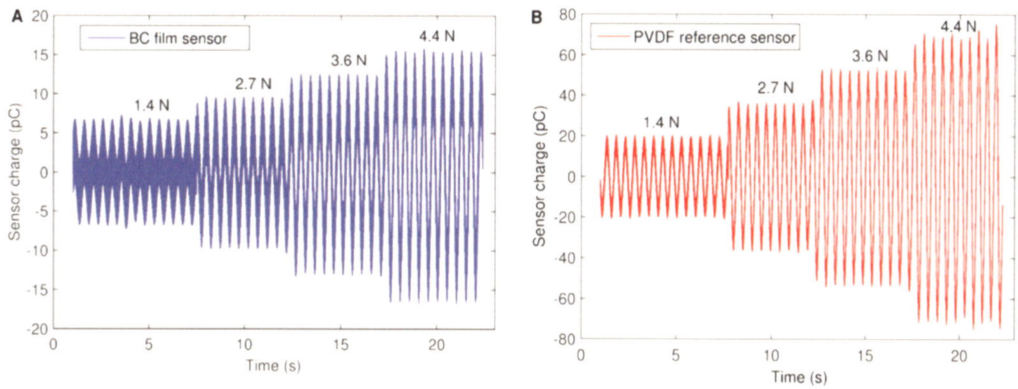

Figure 7. Piezoelectric responses of (a) BC film (produced by WT *K. xylinus*) and (b) commercial PVDF (fabricated using sandwiched PET-Cu electrodes as contacts) sensors under increased mechanical load.

4. Discussion

The SEM images of CNF and BC films revealed a layered structure, resembling that of a ferroelectret material, ElectroMechanical Film (EMFi), widely used in sensor applications [25]. As shown in **Figure 2**, both nanocellulose films have rather smooth surface having a randomly oriented fibrous structure. The density of CNF film (1.38 g/cm^3) is similar to other polymer films, 178 g/cm^3 for PVDF [31] and 0.33 g/cm^3 for EMFi [25].

The sensitivity of wood-based CNF film is around 2–5 pC/N [18, 27]. The sensitivity values measured for CNF sensors slightly vary, but the difference can be mainly explained by the different areas of the sensor films and different electrode materials used. The sensor structure was obtained by sandwiching the CNF film between the copper electrodes (deposited PET film), which may also affect the measured sensitivity values. This is related to the fact that a contact between the sensing material (CNF film) and the electrodes is not as good as in the case of electrodes fabricated directly on the sensing film surface. In comparison, less variation in sensitivity values was observed when electrodes were either screen printed or evaporated directly on PVDF film [23] when compared to PVDF sensors sandwiched between the PET-Cu electrodes.

The CNC sensors showed 2–4 times larger sensitivities than the CNF sensors ((7.3 ± 2.5) pC/N versus ((2.2 ± 1.5) pC/N) [27]. This result was expected since the CNC film consists of solely crystalline cellulose building blocks, whereas CNF contains both crystalline CNCs (cellulose II) domains and amorphous cellulose domains. However, the higher aspect ratio and the stronger ability for entanglement in the case of CNF in comparison with CNC whiskers may offer more versatile options over bare CNC as long as the synthesis of self-standing films and nanopapers is profitable. Despite the intuitively expected purely random alignment of CNCs inside the CNF films, the CNF films may exhibit piezoelectric nature due to the film fabrication process driven non-intentional alignment of CNCs [18].

The sensitivity of BC film varied from (5.0 ± 1.9) pC/N to (16.8 ± 3.9) pC/N, depending on the film type (see **Table 1**) [19]. Overall, the sensitivity values obtained in our approaches for wood-based and bacterial nanocellulose are comparable with values reported by others. For instance, Mahadeva *et al.* reported successful fabrication of hybrid paper from barium titanate nanoparticles and wood cellulose fiber and demonstrated piezoelectric coefficient values of d_{33} = 4.8 pC/N [32]. When compared to commercial piezoelectric materials, quartz has rather low but stable piezoelectricity (d_{33} coefficient of around 2 pC/N) and among the poled polymers PVDF exhibits the highest piezoelectricity (d_{33} coefficient of around 25 pC/N) [33]. However, by improving the nanocellulose film fabrication process, also higher sensitivities for nanocellulose may be obtained. For instance, it has been suggested that regenerated CNC (cellulose II) has the apparent piezoelectric coefficient in the range 35–60 pC/N [15].

The level of crystallinity in cellulose films is considered to have a significant effect on the apparent level of piezoelectricity. Thus, we have measured the crystallinity indices (CI) of BC films using X-ray diffractometer (XRD) [19]. The CI of BC films varied depending on the used bacterial strains (film type) from 88.6 to 97.5%. Similar crystallinity indices were recently reported for wood-based cellulose [32]. The similarity in the XRD profiles reflects the presence of native cellulose I type nanocrystals in bacterial cellulose [34] as well as in plant-based cellulose from wood [32, 35] cotton. The alterations that we observed in piezoelectric sensitivities of CNF films and different BC film types cannot be understood solely based on XRD analysis and there are other factors involved, such as film thickness, film surface roughness, as well as elastic modulus and deformations in the films under applied force.

It should be noted that the nanocellulose films presented in this chapter were not optimized as a piezoelectric material and were not polarized or oriented. However, it cannot be ruled out that the film fabrication processes, including filtering and hot-pressing for CNF and fermentation process for BC, may have caused some alignment of the cellulose nanofibrils inside the thick film. The piezoelectric sensitivities shown here are expected to significantly increase after polarization of the films. In the poling process, an electric field is applied over the processed film for a certain period of time in order to generate piezoelectric properties. The poling (or another type of orientation) of the nanocellulose film is expected to lead into alignment of the crystalline regions inside the film, further leading to a remarkable increase in the piezoelectric effect due to the large dipole momentum of the crystalline components [14, 16]. The initial results of orientation of water processed CNF films, using DC voltage, suggest the potential of remarkably higher sensitivity values [18]. In addition, initial tests with BC films showed that polarization using DC voltage can enhance the piezoelectric sensitivity of the BC sensors. However, polarization of dry film is difficult due to the entangled and stiff structure of the film, and thus, further fabrication process development is required in order to increase the piezoelectric sensitivity of nanocellulose films.

Even though both nanocellulose materials, both wood-based and bacterial nanocellulose, exhibit evident piezoelectric properties, also fundamental differences between these materials exist. First, the fabrication processes of the films are fundamentally different: the wood-based nanocellulose is fabricated from wood cellulose by cutting the fibers into nanoscale units, which can be considered as 'top-down approach', whereas the bacterial cellulose is grown through the oxidative fermentation by bacteria which can be considered as 'bottom-up approach'. Wood-based cellulose has certain limitations, such as multistep mechanical processing to small dimensional constituents resulting in relatively low tensile strength and low flexibility of cellulose films. The most significant restriction in bacterial cellulose production is the slow and relatively expensive fabrication process.

5. Conclusions

In this chapter, we have summarized the recent results obtained with nanocellulose films and compared them with reference piezoelectric polymer film. Both wood-based CNF and CNC films and BC films were discussed here, and the main focus was in the piezoelectricity of these materials. In addition, the results related to microstructure, and dielectric and mechanical properties were discussed.

To conclude, the nanocellulose is an interesting renewable bio-based nanomaterial with potential applications in different fields, such as cell culturing medium, bio-plastic films, high surface-area membranes, or reinforcement material in composites. The more recent application for nanocellulose is its use as bio-based piezoelectric active material. Even though promising results related to the piezoelectricity of the nanocellulose have already been demonstrated, further work to improve the piezoelectric performance of the nanocellulose films is still needed. There are different ways to improve the piezoelectric performance of nanocellulose films by controlled orientation of crystalline areas of nanocellulose films, which will make it suitable for sensors, actuators and energy harvesting applications.

Author details

Sampo Tuukkanen[1]* and Satu Rajala[2]

*Address all correspondence to: sampo.tuukkanen@tut.fi

1 BioMediTech Institute and Faculty of Biomedical Sciences and Engineering, Tampere University of Technology, Tampere, Finland

2 Nokia Technologies, Espoo, Finland

References

[1] Moon RJ, Martini A, Nairn J, Simonsen J, Youngblood J. Cellulose nanomaterials review: Structure, properties and nano-composites. Chemical Society Reviews. 2011;**40**(7):3941-3994. DOI: 10.1039/c0cs00108b

[2] Vuorinen T, Zakrzewski M, Rajala S, Lupo D, Vanhala J, Palovuori K, Tuukkanen S. Printable, transparent, and flexible touch panels working in sunlight and moist environments. Advanced Functional Materials. 2014;**24**(40):6340-6347. DOI: 10.1002/adfm.201401140

[3] Lehtimäki S, Suominen M, Damlin P, Tuukkanen S, Kvarnström C, Lupo D. Preparation of supercapacitors on flexible substrates with electrodeposited PEDOT/graphene composites. ACS Applied Materials & Interfaces. 2015;**7**(40):22137-22147. DOI: 10.1021/acsami.5b05937

[4] Virtanen J, Pammo A, Keskinen J, Sarlin E, Tuukkanen S. Pyrolysed cellulose nanofibrils and dandelion pappus in supercapacitor application. Cellulose. 2017;**24**(8):3387-3397. DOI: 10.1007/s10570-017-1332-8

[5] Rim YS, Bae S-H, Chen H, De Marco N, Yang Y. Recent progress in materials and devices toward printable and flexible sensors. Advanced Materials. 2016;**28**(22):4415-4440. DOI: 10.1002/adma.201505118

[6] Isoniemi T, Tuukkanen S, Cameron DC, Simonen J, Toppari JJ. Measuring optical anisotropy in poly(3,4-ethylene dioxythiophene):poly(styrene sulfonate) films with added graphene. Organic Electronics. 2015;**25**:317-323. DOI: 10.1007/s13391-015-5195-6

[7] Tuukkanen S, Lehtimaki S, Jahangir F, Eskelinen A-P, Lupo D, Franssila S. Printable and disposable supercapacitor from nanocellulose and carbon nanotubes. In: Proceedings of the 5th Electronics System-integration Technology Conference (ESTC). IEEE; 2014. pp. 1-6. DOI: 10.1109/ESTC.2014.6962740

[8] Torvinen K, Lehtimäki S, Keränen JT, Sievänen J, Vartiainen J, Hellén E, Lupo D, Tuukkanen S. Pigment-cellulose nanofibril composite and its application as a separator-substrate in printed supercapacitors. Electronic Materials Letters. 2015;**11**(6):1040-1047. DOI: 10.1007/s13391-015-5195-6

[9] Lehtimäki S, Tuukkanen S, Pörhönen J, Moilanen P, Virtanen J, Honkanen M, Lupo D. Low-cost, solution processable carbon nanotube supercapacitors and their characterization. Applied Physics A: Materials Science & Processing. 2014;**117**(3):1329-1334. DOI: 10.1007/s00339-014-8547-4

[10] Pääkkö M, Ankerfors M, Kosonen H, Nykänen A, Ahola S, Österberg M, Ruokolainen J, Laine J, Larsson PT, Ikkala O, Lindström T. Enzymatic hydrolysis combined with mechanical shearing and high-pressure homogenization for nanoscale cellulose fibrils and strong gels. Biomacromolecules. 2007;**8**(6):1934-1941. DOI: 10.1021/bm061215p

[11] Österberg M, Vartiainen J, Lucenius J, Hippi U, Seppälä J, Serimaa R, Laine J. A fast method to produce strong NFC films as a platform for barrier and functional materials. ACS Applied Materials & Interfaces. 2013;**5**(11):4640-4647. DOI: 10.1021/am401046x

[12] Fukada E. Piezoelectricity of wood. Journal of the Physical Society of Japan. 1955;**10**(2):149-154. DOI: 10.1143/JPSJ.10.149

[13] Fukada E. Piezoelectricity as a fundamental property of wood. Wood Science and Technology. 1968;**2**(4):299-307. DOI: 10.1007/BF00350276

[14] Csoka L, Hoeger IC, Rojas OJ, Peszlen I, Pawlak JJ, Peralta PN. Piezoelectric effect of cellulose nanocrystals thin films. ACS Macro Letters. 2012;**1**(7):867-870. DOI: 10.1021/mz300234a

[15] Cheng H. Flexoelectric nanobiopolymers (FEPs) exhibiting higher mechanical strength (7.5 GPa), modulus (250 GPa), and energy transfer efficiency (75%). Worldwide Electroactive Polymer (Artificial Muscles) Newsletter. 2008;**10**(2):5-7

[16] Frka-Petesic B, Jean B, Heux L. First experimental evidence of a giant permanent electric-dipole moment in cellulose nanocrystals. Europhysics Letters. 2014;**107**(2):28006. DOI: 10.1209/0295-5075/107/28006

[17] Harrison JS, Ounaies Z. Piezoelectric polymers. In: Encyclopedia of Polymer Science and Technology. Hoboken, NJ, USA: John Wiley & Sons, Inc.; 2002. p. 3. DOI: 10.1002/0471440264.pst427

[18] Rajala S, Siponkoski T, Sarlin E, Mettänen M, Vuoriluoto M, Pammo A, Juuti J, Rojas OJ, Franssila S, Tuukkanen S. Cellulose nanofibril film as a piezoelectric sensor material. ACS Applied Materials & Interfaces. 2016;**8**(24):15607-15614. DOI: 10.1021/acsami.6b03597

[19] Mangayil R, Rajala S, Pammo A, Sarlin E, Luo J, Santala V, Karp M, Tuukkanen S. Engineering and characterization of bacterial nanocellulose films as low cost and flexible sensor material. ACS Applied Materials & Interfaces. 2017;**9**(22):19048-19056. DOI: 10.1021/acsami.7b04927

[20] Horiuchi S, Tokura Y. Organic ferroelectrics. Nature Materials. 2008;**7**(5):357-366. DOI: 10.1038/nmat2137

[21] Kärki S, Lekkala J, Kuokkanen H, Halttunen J. Development of a piezoelectric polymer film sensor for plantar normal and shear stress measurements. Sensors and Actuators A: Physical. 2009;**154**(1):57-64. DOI: 10.1016/j.sna.2009.07.010

[22] Rajala S, Salpavaara T, Tuukkanen S. Testing and comparing of film-type sensor materials in measurement of plantar pressure distribution. In: Proceedings of 38th Annual International Conference of the IEEE Engineering in Medicine and Biology Society (EMBC); 17-20 August 2016; Orlando, FL, USA. DOI: 10.1109/EMBC.2016.7590687

[23] Rajala S, Tuukkanen S, Halttunen J. Characteristics of piezoelectric polymer film sensors with solution-processable graphene-based electrode materials. IEEE Sensors Journal. 2015;**15**(6):3102-3109. DOI: 10.1109/JSEN.2014.2344132

[24] Rajala S, Mettänen M, Tuukkanen S. Structural and electrical characterization of solution-processed electrodes for piezoelectric polymer film sensors. IEEE Sensors Journal. 2016;**16**(6):1692-1699. DOI: 10.1109/JSEN.2015.2504956

[25] Rajala S, Paajanen M, Lekkala J. Measurement of sensitivity distribution map of a ferroelectret polymer film. IEEE Sensors Journal. 2016;**16**(23):8517-8522. DOI: 10.1109/JSEN.2016.2613876

[26] Rajala S, Vuoriluoto M, Rojas OJ, Franssila S, Tuukkanen S. Piezoelectric sensitivity measurements of cellulose nanofibril sensors. In: Proceedings of the XXI IMEKO World Congress; Prague, Czech Republic. 2015. pp. 2-6

[27] Tuukkanen S, Rajala S. A survey of printable piezoelectric sensors. In: Proceedings of the IEEE SENSORS Conference. Busan, South Korea: IEEE, 2015. pp. 1-4. DOI: 10.1109/ICSENS.2015.7370542

[28] Hänninen A, Rajala S, Salpavaara T, Kellomäki M, Tuukkanen S. Piezoelectric sensitivity of a layered film of chitosan and cellulose nanocrystals. Procedia Engineering. 2016;**168**:1176-1179. DOI: 10.1016/j.proeng.2016.11.397

[29] Ramadan KS, Sameoto D, Evoy S. A review of piezoelectric polymers as functional materials for electromechanical transducers. Smart Materials and Structures. 2014;**23**(3):033001. DOI: 10.1088/0964-1726/23/3/033001

[30] Fraden J. Handbook of Modern Sensors: Physics, Designs, and Applications. 2nd ed. New York: Springer-Verlag; 2010. p. 663. DOI: 10.1007/978-1-4419-6466-3

[31] Measurement Specialties Inc. Piezo film sensors, Technical manual [Internet]. Available from: http://www.meas-spec.com [Accessed: May 5, 2014]

[32] Mahadeva SK, Walus K, Stoeber B. Piezoelectric paper fabricated via nanostructured barium titanate functionalization of wood cellulose fibers. ACS Applied Materials & Interfaces. 2014;**6**(10):7547-7553. DOI: 10.1021/am5008968

[33] Regtien P, Van Der Heijden F, Korsten MJ, Otthius W. Measurement Science for Engineers. London: Kogan Page Science; 2004. p. 358

[34] Feng X, Ullah N, Wang X, Sun X, Li C, Bai Y, Chen L, Li Z. Characterization of bacterial cellulose by Gluconacetobacter hansenii CGMCC 3917. Journal of Food Science. 2015; **80**(10):E2217-E2227. DOI: 10.1111/1750-3841.13010

[35] Mahadeva SK, Kim J. Hybrid nanocomposite based on cellulose and tin oxide: Growth, structure, tensile and electrical characteristics. Science and Technology of Advanced Materials. 2011;**12**(5):055006. DOI: 10.1088/1468-6996/12/5/055006

Chapter 2

Application of Thin Piezoelectric Films in Diamond-Based Acoustoelectronic Devices

Boris P. Sorokin, Gennady M. Kvashnin,
Andrey S. Novoselov, Sergey I. Burkov,
Anton B. Shipilov, Nikolay V. Luparev,
Victor V. Aksenenkov and Vladimir D. Blank

Additional information is available at the end of the chapter

http://dx.doi.org/10.5772/intechopen.76715

Abstract

The theory of external loading influence on acoustic parameters of piezoelectric five-layered structure as "Al/(001) AlN/Mo/(001) diamond/Me" has been developed. Oscillations in diamond-based high-overtone bulk acoustic resonators (HBARs) have been investigated in terms of 3D FEM simulation. Peculiarities of technology of aluminum-scandium nitride (ASN) films have been discussed. Composition $Al_{0.8}Sc_{0.2}N$ was obtained to create the diamond-based HBAR and SAW resonator. Application of ASN films has resulted in a drastic increasing an electromechanical coupling up to 2.5 times in comparison with aluminum nitride. Development of ASN technology in a way of producing a number of compositions with the better piezoelectric properties has a clear prospective. SAW resonator based on "Al IDT/(001) AlN/(001) diamond" structure has been investigated in the band 400–1500 MHz. The highest-quality factor $Q \approx 1050$ was observed for the Sezawa mode at 1412 MHz. Method of measuring HBAR's parameters within 4–400 K at 0.5–5 GHz has been developed. Results on temperature dependence of diamond's Q-factor at relatively low frequencies were quite different in comparison with the ones at the frequencies up to 5 GHz. Difference could be explained in terms of changing mechanism of acoustic attenuation from Akhiezer's type to the Landau-Rumer's one at higher frequencies in diamond.

Keywords: aluminum nitride, aluminum-scandium nitride, diamond-based HBAR, diamond-based SAW resonator, acoustoelectronic sensor, UHF

1. Introduction

Thin piezoelectric films (TPF) are the important components of many electronic sandwich devices due to its compatibility with the planar technology and comparatively low cost. Combination of effective TPF with non-piezoelectric substrate possessing the good acoustic properties allows obtaining the new micro devices especially in UHF bands. The matter is that the known single-crystalline piezoelectric materials as quartz, lithium niobate, tantalate, langasite, and langatate do not own a set of useful properties such as high electromechanical coupling, thermostable cuts, and good acoustic properties joined within a given crystal. Additionally, an enhancement of operational frequencies up to several GHz for bulk acoustic wave (BAW) resonators requires the thinning of crystal plates to the thickness close to several microns which seems really impossible because a lot of defects will be arisen. So, producing the best quality piezoelectric films compatible with any substrate materials is one of the main tasks of modern piezoelectric technology.

Since the end of the twentieth century, the zinc oxide (ZnO) and aluminum nitride (AlN) polycrystalline films with the wurtzite structure were widely used in such acoustoelectronic devices as composite resonators, filters, duplexers, etc. [1, 2]. It seems that the aluminum nitride has a more preferable combination of required qualities due to high dielectric properties as well as good temperature stability up to 1000 °C. First, in 2009 the authors [3] have found an enhancement of piezoelectric response in aluminum-scandium nitride (ASN) $Al_{1-x}Sc_xN$ solid solution. Therefore, aluminum-scandium nitride should be derived as prospective piezoelectric material for acoustoelectronic devices including sensors and energy harvesters. But the electromechanical properties of aluminum-scandium nitride are strongly depended on the scandium content, and as a result, its application in acoustoelectronic devices is discussible now.

High-overtone bulk acoustic resonator (HBAR) differs from conventional piezoelectric resonators due to their small size and high-quality factor Q at microwave operational frequencies [1]. A lot of materials have been successfully used to prepare the HBAR substrates: crystalline quartz, fused silica, silicon [4], sapphire [4, 5], and yttrium aluminum garnet (YAG) [6]. But the highest frequencies of the acoustic overtone excitation did not exceed 10 GHz. Recently, a distinctive progress in the achievement of record resonance frequencies up to 20 GHz has been emphasized by the development of HBAR based on piezoelectric layered structure (PLS) involving the single-crystalline diamond substrate and AlN thin-film piezoelectric transducer (TFPT) [7].

Naturally, thin piezoelectric films are widely used as the important elements in acoustoelectronic devices on surface acoustic waves (SAW), such as SAW delay lines, resonators, filters, and sensors. For example, in Ref. [8] the one-port SAW resonator, based on the layered structure AlN/polycrystalline diamond/Si and excited on the resonant frequency ~1.35 GHz, has been described. SAW filter at the frequency band ~6.3 GHz based on the structure SiO_2/AlN/polycrystalline diamond has been developed by authors [9].

A main aim of this paper was defined by the necessity in describing the modern trends in application of thin piezoelectric films as active elements in microwave acoustoelectronic composite devices, including the choice and development of the new effective piezoelectric materials compatible with single-crystalline diamond substrates.

2. Theory of the thin-film loading influence on the acoustic parameters of diamond-based piezoelectric layered structure

An effective application of multilayered piezoelectric structure as a prototype of specialized acoustoelectronic sensor device should be based on the theory of the influence of an external loading on its acoustic parameters.

Propagation of the small amplitude acoustic waves of a piezoelectric crystal could be described basically by the equations of electrostatics and the equations of state of a piezoelectric medium written in coordinate form [10]:

$$\begin{aligned} \rho_0 \ddot{U}_i = \tau_{ij,j}; \quad D_{m,m} = 0; \\ \tau_{ij} = C^E_{ijkl}\eta_{kl} - e_{mij}E_m; \quad D_m = \varepsilon^{\eta}_{mn}E_n + e_{mij}\eta_{ij}, \end{aligned} \tag{1}$$

where ρ_0 is the density of a crystal, $\mathbf{U}_i$ is the vector of dynamic elastic displacements, τ_{ij} is the tensor of thermodynamic stresses, $\mathbf{D}_m$ is the vector of electrical induction, η_{kl} is the tensor of small deformations, C^E_{ijkl}, e_{mij}, and ε^{η}_{mn} are second-order elastic, piezoelectric, and clamped dielectric constants, respectively. The comma after the subscript denotes a spatial derivative, and coordinate Latin indices vary from 1 to 3. Here and further, the rule of summation over repeated indices will be used. For elastic displacements and the electric potential in the form of small-amplitude plane monochromatic waves, the system of equations was written in the form of the well-known Green-Christoffel equations, which must be solved for each layered medium to be used:

$$\begin{aligned} \left(\Gamma_{ik} - \rho_0\omega^2\delta_{ik}\right)\alpha_i = 0; \qquad i,k = 1\ldots4; \qquad \delta_{44} = 0; \\ \Gamma_{ik} = C_{ijkm}k_jk_m; \ \Gamma_{4j} = \Gamma_{j4} = e_{ijk}k_ik_k; \ \Gamma_{44} = -\varepsilon^{\eta}_{nm}k_nk_m \end{aligned} \tag{2}$$

where $\vec{k} = \frac{\omega}{V}\vec{n}$ is the wave vector; $\vec{n}$ is the unit vector of wave propagation; ω and V are the angular frequency and phase velocity of an elastic wave, respectively; α_i is the component of eigenvectors of elastic displacement; and α_4 is the amplitude of the wave of quasistatic electrical potential connected with an elastic wave. Solving of Green-Christoffel equations taken in a general form, Eq. (2) allows to define the parameters of propagation both the bulk and surface acoustic waves in piezoelectric layered structure.

Let the X_3 axis of operational coordinate system be directed along the outer normal to the surface of the layer, and the X_1 axis coincides with the wave propagation direction (**Figure 1**). In a multilayer piezoelectric structure, describing the propagation of the elastic waves, it is necessary to write down the special boundary conditions, which, depending on the number m of layers, take a form:

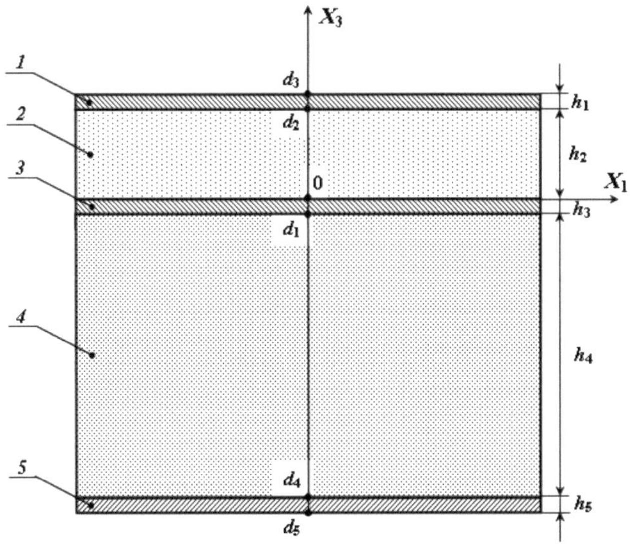

Figure 1. Arrangement of a piezoelectric layered structure: (*1*) top electrode, (*2*) piezoelectric film, (*3*) bottom electrode, (*4*) substrate, and (*5*) the layer to be investigated.

$$
\begin{aligned}
&\tau_{3i}^{(1)} = 0\Big|_{x_3=h_1}, \quad D_3^{(1)} = D^{(вак)}\Big|_{x_3=h_1};\\
&\tau_{3i}^{(1)} = \tau_{3i}^{(2)}\Big|_{x_3=h_2}, \quad D_3^{(1)} = D_3^{(2)}\Big|_{x_3=h_2}; \varphi^{(1)} = \varphi^{(2)}\big|_{x_3=h_2}; \quad U_i^{(1)} = U_i^{(2)}\Big|_{x_3=h_2};\\
&\cdots\cdots\cdots\\
&\tau_{3i}^{(m-1)} = \tau_{3i}^{(m)}\Big|_{x_3=0}, \quad D_3^{(m-1)} = D_3^{(m)}\Big|_{x_3=0}; \varphi^{(m-1)} = \varphi^{(m)}\big|_{x_3=0}; \quad U_i^{(m-1)} = U_i^{(m)}\Big|_{x_3=0};\\
&\tau_{3i}^{(m)} = 0\Big|_{x_3=-h_n}, \quad D_3^{(m)} = D^{(вак)}\Big|_{x_3=-h_m}.
\end{aligned}
\tag{3}
$$

Here, h_1, …, h_m are the layer thicknesses. Substituting into boundary conditions (3) the required solutions as a linear combinations of the partial waves

$$
\begin{aligned}
U_i^{(m)} &= \sum_n C_n^{(m)} \alpha_i^{(n)} \exp\left[i\left(k_1 x_1 + k_3{}^{(n)} x_3 - \omega t\right)\right],\\
\Phi^{(m)} &= \sum_n C_4^{(m)} \alpha_4^{(n)} \exp\left[i\left(k_1 x_1 + k_3{}^{(n)} x_3 - \omega t\right)\right],
\end{aligned}
\tag{4}
$$

one can obtain a matrix of boundary conditions, and the vanishing of its determinant allows obtaining the equations for determining the parameters of elastic wave propagation. In Eq. (4) a superscript n is associated with nth partial wave. By varying the boundary conditions (Eq. (3)), one can specify all the types of PLS elastic waves. For example, the first equation in Eq. (3) is related to the Rayleigh-type SAW propagation. The first and last equations together describe the propagation of Lamb-type elastic waves in a piezoelectric plate with a thickness h, i.e., in this case Eq. (3) at m = 1. taking the form

$$\begin{cases} \sum_{n=1}^{8} C_n \left[C_{3jkm} k_m^{(n)} \alpha_k^{(n)} + e_{k3j} k_k^{(n)} \alpha_4^{(n)} \right] \exp\left[i k_3^{(n)} h \right] = 0; \\ \sum_{n=1}^{8} C_n \left[e_{3km} k_m^{(n)} \alpha_k^{(n)} - \left(\varepsilon_{3k} k_k^{(n)} - i\varepsilon_0 \right) \alpha_4^{(n)} \right] \exp\left[i k_3^{(n)} h \right] = 0; \\ \sum_{n=1}^{8} C_n \left[C_{3jkm} k_m^{(n)} \alpha_k^{(n)} + e_{k3j} k_k^{(n)} \alpha_4^{(n)} \right] = 0; \\ \sum_{n=1}^{8} C_n \left[e_{3km} k_m^{(n)} \alpha_k^{(n)} - \left(\varepsilon_{3k} k_k^{(n)} + i\varepsilon_0 \right) \alpha_4^{(n)} \right] = 0. \end{cases} \tag{5}$$

At the assumption that the lower layer is sufficiently thick (semi-infinite), i.e., the thickness of a layer should be much greater than the length of the elastic wave, in which case the last equation in Eq. (3) cannot be taken into account, i.e., the presence of a free lower boundary will be ignored. It is also necessary to require that the condition $\mathrm{Im}\left(k_3{}^{(m)}\right) < 0$ ensuring the attenuation of the elastic wave into the depth of a substrate should be satisfied. Thus, the boundary conditions (Eq. (3)) describing the elastic wave propagation in the "layer-substrate" structure will have the form.

$$\begin{gathered} \sum_{n=1}^{8} \left[a_n C_{3kpi}^{(1)} k_i^{(n)} \alpha_p^{(n)} + a_4 e_{p3k}^{(1)} k_p^{(n)} \alpha_4^{(n)} \right] \exp\left[i k_3^{(n)} h \right] = 0, \\ \sum_{n=1}^{8} \left[a_n e_{3kl}^{(1)} k_l^{(n)} \alpha_k^{(n)} + a_4 \left(\varepsilon_{3k}^{(1)} k_k^{(n)} - i\varepsilon_0 \right) \alpha_4^{(n)} \right] \exp\left[i k_3^{(n)} h \right] = 0, \\ \sum_{m=1}^{4} b_m \left[C_{i3kl}^{(2)} k_l^{(s)} \alpha_k^{(s)} + e_{3pi}^{(2)} k_p^{(s)} \alpha_4^{(s)} \right] - \sum_{n=1}^{8} a_n \left[C_{i3kl}^{(1)} k_l^{(n)} \alpha_k^{(n)} + e_{3pi}^{(1)} k_p^{(n)} \alpha_4^{(n)} \right] = 0, \\ \sum_{m=1}^{4} b_m \left[e_{3kl}^{(2)} k_l^{(s)} \alpha_k^{(s)} + \varepsilon_{3k}^{(2)} k_k^{(s)} \alpha_4^{(s)} \right] - \sum_{n=1}^{8} a_n \left[e_{3kl}^{(1)} k_l^{(n)} \alpha_k^{(n)} + \varepsilon_{3k}^{(1)} k_k^{(n)} \alpha_4^{(n)} \right] = 0, \\ \sum_{s=1}^{4} U_i^{(2)(s)} b_m - \sum_{n=1}^{8} U_i^{(1)(n)} a_n = 0. \end{gathered} \tag{6}$$

Here, the digital superscripts 1 and 2 denote the layer and the substrate, respectively; $\alpha_k^{(s)}$ and b_m are the amplitude and weight coefficients of the mth partial wave (m = 1, …, 4) in a substrate; and $\alpha_k^{(n)}$ and a_n are the amplitude and weight coefficients of the nth partial wave (n = 1, …, 8) in a piezoelectric layer.

In modern acoustoelectronics, accurate information about the mechanical parameters of thin layers of new materials and thin mono- and polycrystalline films used to create microwave acoustic resonators, filters, and sensors should be of great importance. Previously proposed and used in a number of experiments, the original method of resonance acoustic microwave spectroscopy has opened a possibility of measuring these parameters [11]. The essence of such method was based on the investigated film that was included into a content of an acoustic composite microwave resonator changing its acoustic parameters (**Figure 1**). Connection with the external electrical circuit was carried out by the TFPT. In the case of thin films, the last one should be deposited on a sufficiently thick substrate made of a material with low acoustic losses. Information on the attenuation coefficient and sound velocity in the film were found from the comparison of the measured total losses and the positions of the resonant peaks without the film and after its deposition.

To solve the problems of acoustic microwave spectroscopy, it is necessary to introduce a system of boundary conditions for the PLS according to **Figure 1**. For elastic displacements in the form of a plane sinusoidal wave, one can be written as

$$U^{(S)} = a^{(S)}\exp\left(-ik^{(S)}x_3\right) + b^{(S)}\exp\left(ik^{(S)}x_3\right), \tag{7}$$

where the subscript S denotes the medium used in calculations; $a^{(S)}$ and $b^{(S)}$ are the amplitudes of the incident and reflected elastic waves in the Sth layer, respectively; and k is the modulus of the wave vector in a layer.

Assuming that the electrodes were consisted of isotropic metals, the piezoelectric layer had the crystalline symmetry as 6 *mm*, the substrate was taken as a cubic dielectric crystal, and the layer to be investigated was isotropic, the boundary conditions have been formulated as follows:

(1) Stress tensor should have the zero normal components at the interface "top electrode-vacuum" in the form

$$\tau_{3j}^{(1)} = 0\Big|_{x_3=d_3} = -a^{(1)}k^{(1)}C^{(1)}\exp\left(-ik^{(1)}d_3\right) + b^{(1)}k^{(1)}C^{(1)}\exp\left(ik^{(1)}d_3\right) = 0, \tag{8}$$

where $k^{(1)}$ is the wave vector at the top electrode layer, $C^{(1)} = C_{11}$ is the elastic modulus of the top electrode film *1*, and d_3 is the point on the interface "top electrode-vacuum." Here and after the superscript designates the PLS layer's number.

(2) Normal components of the stress tensor and displacement vectors should be equal to appropriate values taken at the interface "top electrode-piezoelectric film" as follows:

$$\begin{gathered}
\tau_{3j}^{(1)} = \tau_{3j}^{(2)}\Big|_{x_3=d_2}, \quad U^{(1)} = U^{(2)}, \\
-a^{(1)}k^{(1)}C^{(1)}\exp\left(-ik^{(1)}d_2\right) + b^{(1)}k^{(1)}C^{(1)}\exp\left(ik^{(1)}d_2\right) - a^{(2)}k^{(2)}C^{(2)}P - b^{(2)}k^{(2)}C^{(2)}P^* = e\frac{V_0}{h_2}, \\
P = \exp\left(-ik^{(2)}d_2\right) + \frac{ie^2}{\varepsilon C^{(2)}k^{(2)}d_2}\left[\exp\left(ik^{(2)}d_2\right) - 1\right], \\
a^{(1)}\exp\left(-ik^{(1)}d_2\right) + b^{(1)}\exp\left(ik^{(1)}d_2\right) - a^{(2)}\exp\left(-ik^{(2)}d_2\right) - b^{(2)}\exp\left(ik^{(2)}d_2\right) = 0.
\end{gathered} \tag{9}$$

In Eq. (9) the following notations are introduced: P^* is a complex conjugated value; $C^{(2)} = C_{33}$, $e = e_{33}$, and $k^{(2)}$ are the elastic modulus, the piezoelectric constant, and the wave vector in piezoelectric film *2*, respectively; d_2 is the point on the interface "top electrode-piezoelectric

http://dx.doi.org/10.5772/intechopen.76715

film"; $U^{(2)}$ and $U^{(1)}$ are the elastic displacement vectors in piezoelectric layer and top electrode, respectively; V_0 is the potential difference between the electrodes; h_2 is the thickness of piezoelectric layer.

(3) Normal components of the stress tensor and displacement vectors should be equal to appropriate values taken at the interface "piezoelectric film-bottom electrode" as follows:

$$\tau_{3j}^{(2)} = \tau_{3j}^{(3)}\Big|_{x_3=0}, \quad U^{(2)} = U^{(3)},$$
$$a^{(3)}k^{(3)}C^{(3)} + b^{(3)}k^{(3)}C^{(3)} + a^{(2)}k^{(2)}C^{(2)} - b^{(2)}k^{(2)}C^{(2)} = e\frac{V_0}{h_2}, \tag{10}$$
$$a^{(3)} + b^{(3)} - a^{(2)} - b^{(2)} = 0.$$

In Eq. (10) the following notations are introduced: $C^{(3)} = C_{11}$, $U^{(3)}$, and $k^{(3)}$ are the elastic modulus, the elastic displacement, and the wave vector in the bottom electrode 3, respectively.

(4) Normal components of the stress tensor and displacement vectors should be equal to appropriate values taken at the interface "bottom electrode-substrate" as follows:

$$\tau_{3j}^{(4)} = \tau_{3j}^{(3)}\Big|_{x_3=d_1}, U^{(4)} = U^{(3)},$$
$$a^{(3)}k^{(3)}C^{(3)}\exp\left(ik^{(3)}d_1\right) - b^{(3)}k^{(3)}C^{(3)}\exp\left(-ik^{(3)}d_1\right) - a^{(4)}k^{(4)}C^{(4)}\exp\left(ik^{(4)}d_1\right) - b^{(4)}k^{(4)}C^{(4)}\exp\left(-ik^{(4)}d_1\right) = 0,$$
$$a^{(3)}\exp\left(ik^{(3)}d_1\right) + b^{(3)}\exp\left(-ik^{(3)}d_1\right) - a^{(4)}\exp\left(ik^{(4)}d_1\right) - b^{(4)}\exp\left(-ik^{(4)}d_1\right) = 0. \tag{11}$$

In Eq. (11) the following notations are introduced: $C^{(4)} = C_{11}$, $U^{(4)}$, and $k^{(4)}$ are the elastic modulus, the elastic displacement, and the wave vector in the substrate 4, respectively, and d_1 is the point on the interface "bottom electrode-substrate."

(5) Normal components of the stress tensor and displacement vectors should be equal to appropriate values taken at the interface "substrate-the layer to be investigated" as follows:

$$\tau_{3j}^{(4)} = \tau_{3j}^{(5)}\Big|_{x_3=d_4}, \quad U^{(4)} = U^{(5)},$$
$$a^{(5)}k^{(5)}C^{(5)}\exp\left(ik^{(5)}d_4\right) - b^{(5)}k^{(5)}C^{(5)}\exp\left(-ik^{(5)}d_4\right) - a^{(4)}k^{(4)}C^{(4)}\exp\left(ik^{(4)}d_4\right) + b^{(4)}k^{(4)}C^{(4)}\exp\left(-ik^{(4)}d_4\right) = 0,$$
$$a^{(5)}\exp\left(ik^{(5)}d_4\right) + b^{(5)}\exp\left(-ik^{(5)}d_4\right) - a^{(4)}\exp\left(ik^{(4)}d_4\right) - b^{(4)}\exp\left(-ik^{(4)}d_4\right) = 0. \tag{12}$$

In Eq. (12) the following notations are introduced: $C^{(5)} = C_{11}$, $U^{(5)}$, and $k^{(5)}$ are the elastic modulus, the elastic displacement, and the wave vector in the layer 5, respectively, and d_4 is the point on the interface "substrate-the layer to be investigated."

(6) Normal components of the stress tensor should be equal to appropriate values taken at the "the layer to be investigated-vacuum" as follows:

$$\begin{aligned} &\tau_{3j}^{(5)} = 0\Big|_{x_3=d_5}, \\ &a^{(5)}k^{(5)}C^{(5)}\exp\left(ik^{(5)}d_5\right) - b^{(5)}k^{(5)}C^{(5)}\exp\left(-ik^{(5)}d_5\right) = 0, \end{aligned} \tag{13}$$

where d_5 is the point on the interface "the layer to be investigated-vacuum."

Boundary conditions (Eq. (8)–(13)) form a system of equations as 10×10 dimensions, the solution of which makes it possible to determine the amplitudes of the elastic waves in all the layers and the PLS frequency characteristics. Note that the part of the boundary condition matrix as 7×7 dimensions allows to obtain the so-called form factor of HBAR including into calculation a traveling acoustic wave only, i.e., without taking into account the lower boundary of the crystalline substrate (12) and (13). Additionally, when Eq. (12) was taken into account, as a result appropriate boundary condition matrix as 8×8 dimensions can be formed, and the HBAR's form factor can be calculated including the influence of a bottom boundary of a substrate.

In order to study the influence of the fifth layer on the PLS acoustic parameters, a required own software "Modeling of the processes of resonant acoustic spectroscopy in multilayered structures" based on the above theory has been developed [12]. An estimation of the influence of metal film deposition as a fifth layer on the change of the overtone resonant frequency of diamond-based PLS "Al/(001) AlN/Mo/(001) diamond/Al (Mo)" is presented in **Figure 2**. In calculations the thickness of Al or Mo films was varied within 0–1300 nm. As one can see, the metal deposition leads to the decrease of resonant frequency with the thickness increasing both the metals. But on the curve associated with Al influence, the sharp variation of resonant frequency in the vicinity of 400–600 nm takes place on the contrary with the linearly

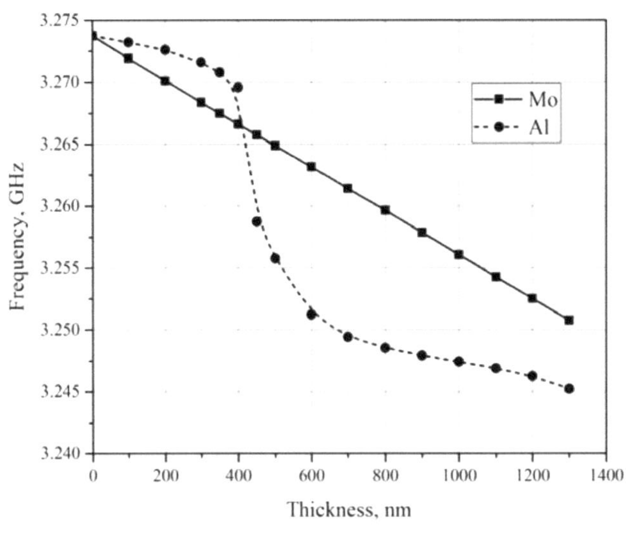

Figure 2. Change in an overtone resonant frequency of diamond-based PLS by metal film deposition. In calculation the PLS "Al/(001) AlN/Mo/(001) diamond/Al (Mo)" was used with thicknesses (in microns) of the films and substrate as "0.14/1.125/0.16/501/0–1.2," respectively.

proportional dependence of resonant frequency on the thickness of Mo film. This can be explained by the quite different acoustic impedances $Z = \rho V$ of Al and Mo as 17 and 63 (in 10^6 kg/s·m^2), respectively. As a result, the elastic waves falling from the substrate on the diamond-Al or diamond-Mo boundaries should be mainly reflected or should pass through the boundary freely, respectively. But because the quarter-wave, thickness in the Al layer is approximately equal to 450 nm at 3.26 GHz; in this case the Al film should become an antireflective one, which leads to a sharp increase in the effective thickness of the substrate and, correspondingly, to a stepwise decrease in the resonant frequency. On the other hand, a monotonic increase in the Mo film thickness leads only to a smooth increase in the effective thickness of the substrate and, correspondingly, to a linearly proportional decrease of the resonant frequency.

3. 3D simulation of acoustic wave propagation in multilayered piezoelectric structure

Earlier [13, 14], we have successfully applied the 2D FEM simulation in order to obtain a quite complex pattern of dispersive dependences of phase velocities of plate Lamb waves observing visualization of the fields of elastic displacements belonging to a lot of acoustic modes with a number of eigenfrequencies. Besides of Lamb waves, the BAW and SAW modes of Rayleigh type were found. But the statement of the task in 2D approximation did not allow obtaining the results on the SH modes.

In the 3D FEM simulation as a model, the PLS "Al/(001) AlN/Mo/(001) diamond" has been investigated by the software COMSOL Multiphysics. Boundary conditions on the top and bottom surfaces of a model sample were chosen as the free ones, and on all the vertical surfaces as the symmetrical ones. Width×length×thickness (in microns) of diamond substrate and AlN film were 1000×1000×392 and 400×400×0.624, respectively. Thicknesses (in nm) of Al and Mo films were taken as 164 and 169, respectively. All the dimensions of a model sample were close to appropriate ones in the experimental sample. In **Figure 3**, the results on the second overtone of longitudinal bulk acoustic wave in the PLS investigated are presented as 2D and 3D images. Note

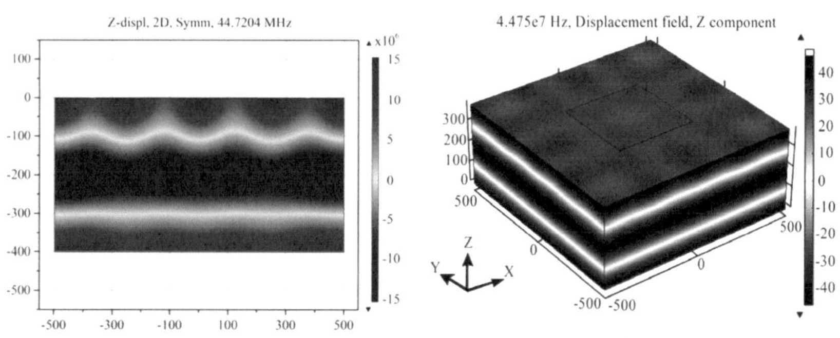

Figure 3. Elastic displacement fields (Z-component) within a model PLS in 2D (a) and 3D (b) approximation for longitudinal bulk acoustic wave propagating along Z-axis.

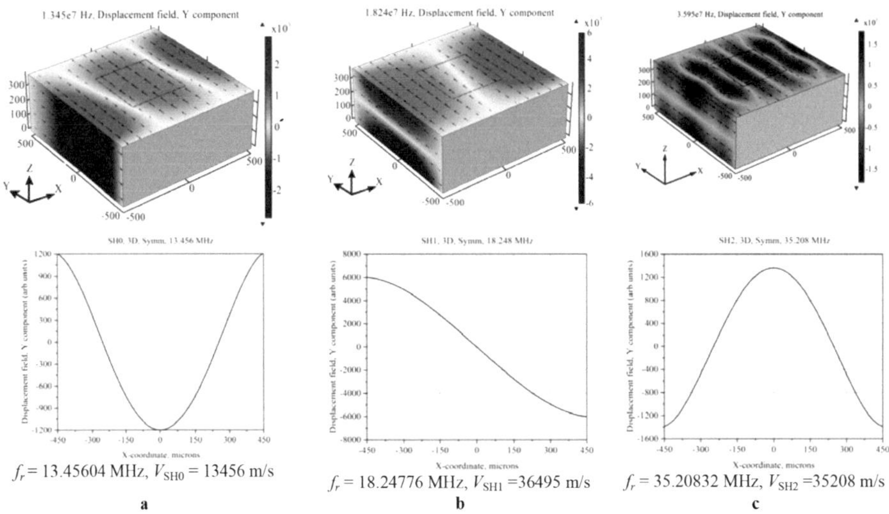

Figure 4. Elastic displacement fields (Y-component) within a model PLS in 3D approximation for SH_0 (a), SH_1 (b), and SH_2 (c) dispersive modes. Arrows should be associated with the vectors of a local elastic displacement.

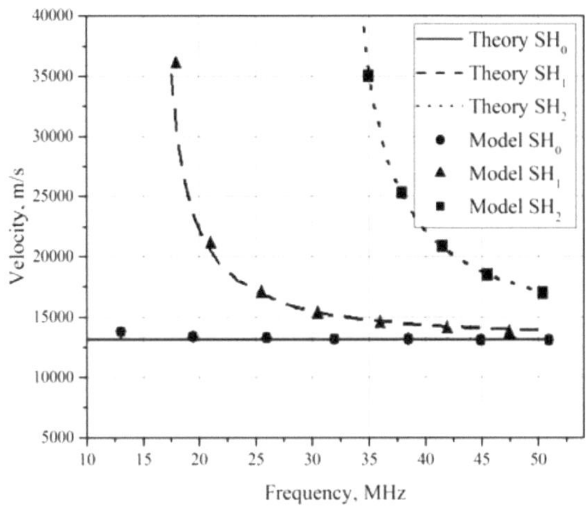

Figure 5. Dependences of SH-type wave velocities on the frequency for the SH_0, SH_1, and SH_2 dispersive branches in the PLS "Al/(001) AlN/Mo/(001) diamond" in comparison with theory data.

that the calculated data on resonant frequency (in MHz) as 44.72 obtained in 2D approximation (**Figure 3a**) are in close accordance with the similar as 44.75 (**Figure 3b**) in 3D approximation.

In **Figure 4**, an example of the data on the elastic displacement patterns arising for the SH-type waves in the PLS "Al/(001) AlN/Mo/(001) diamond" is presented in 3D image. SH waves are dispersive and deeply penetrating into a substrate. As a consequence, SH wave velocities should be strongly depended on the operational frequency. Such dependences for a number of first

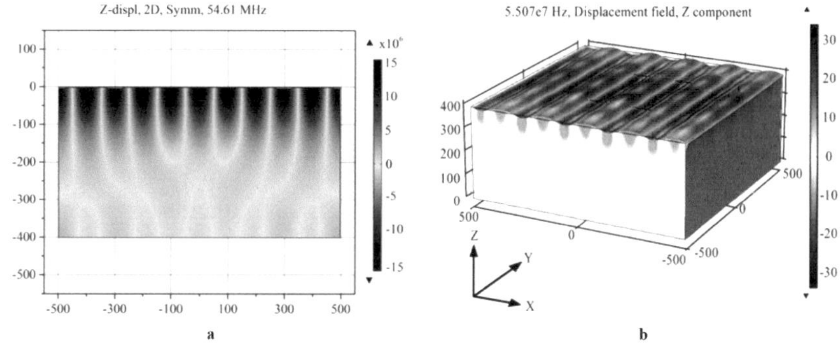

Figure 6. Z-components of elastic displacement fields within a model PLS "Al/(001) AlN/Mo/(001) diamond" in 2D (a) and 3D (b) approximation for Rayleigh-type surface wave propagating along X-axis.

dispersive branches within the frequency band 10–50 MHz are presented in **Figure 5** in comparison with theory data. As one can see, a good agreement between the model and theory was obtained.

In **Figure 6** an example of the elastic displacement patterns arising for the Rayleigh-type waves in the PLS "Al/(001) AlN/Mo/(001) diamond" is presented in 2D and 3D image. As one can calculate, the SAW wave lengths in a diamond were both equal to 200 microns in 2D and 3D approximation, and the SAW phase velocities were equal to 10,922 and 11,014 m/s, respectively. One can speak about a reasonable agreement between the results of two models.

4. Aluminum nitride and aluminum-scandium nitride film preparation: sample characterization

In our microwave experiments, a lot of devices under test (DUT) operating the BAW or SAW propagation in PLS were required (**Figure 7**). As one can see, it was necessary to fabricate the

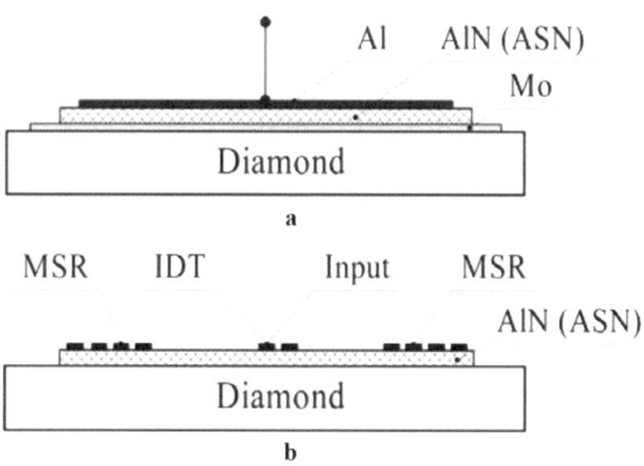

Figure 7. Devices under test. (a) BAW resonator based on the PLS "Al/AlN or ASN piezoelectric film/Mo/diamond" and (b) single-port SAW resonator based on the PLS "Al IDT/AlN or ASN piezoelectric film/diamond." IDT and MSR are the interdigital transducer and microstrip reflector, respectively.

multilayered piezoelectric structures involving the TFPT and diamond substrate (**Figure 7a)** or IDT/AlN (ASN) piezoelectric film on diamond substrate (**Figure 7b**). Here, IDT is an interdigital transducer conventionally applied in a number of SAW devices to SAW exciting and detection.

The IIa type synthetic diamond single crystals grown by HPHT method at the Technological Institute for Superhard and Novel Carbon Materials were used as substrates for producing in all the DUTs investigated. Such crystals have the good dielectric properties and a low content of nitrogen impurity. All the substrates with the orientation of the main surfaces (100), (110), and (111) were double side polished up to roughness R_a within 0.2–2 nm on 10 × 10 μm^2 area controlled by AFM method. As a result of the mechanical grinding and polishing, roughness and a buried damage layer (BDL) underneath the crystalline surface should take place. In order to control the BDL depth, we applied the method of electron backscatter diffraction (EBSD) and Kikuchi line (KL) observation [15]. Appearance of clear KL is closely correlated with the BDL absence. As a summary, one can highlight the unique peculiarity of single-crystalline diamond: multiple EBSD experiments involving the faces with different treatment degrees, as well as natural faces of as-grown crystal, have produced practically the same clear KL patterns. So, the BDL depth of the diamond substrate can be estimated as several atomic layers. This fact was first discovered by us and drastically singles out crystalline diamond from the set of any other conventionally applications in acoustoelectronic device materials. As a result, the BDL can only have a weak influence on the UHF acoustic properties of diamond.

Metal electrodes, aluminum nitride or aluminum-scandium nitride piezoelectric films, were deposited by magnetron sputtering equipment AJA ORION 8. Process of aluminum-scandium nitride synthesis in comparison with that for AlN was differed by the application of additional Sc target, so the Al and Sc targets were in work simultaneously. One can see from the available experimental data [16] that a significant increase in piezoelectric module d_{33} (more than in four times) in solid solution $Al_{1-x}Sc_xN$ has took place with increasing scandium concentration. On the other hand, pure scandium nitride (ScN) should belong to a cubic symmetry with the centrosymmetrical space group Fm3m and the rock salt structure (*rs*), which leads to an absence of piezoelectricity. It was shown [16] that the hexagonal piezoelectric structure of wurtzite type (*wz*) has been stably existed within $0 \leq x \leq 0.43$ only. Coexistence of *wz* and *rs* phases should be observed at $0.43 \leq x \leq 0.55$, while the *rs* structure was preferably observed when $0.55 < x \leq 1.0$. So, a required Sc concentration leading to the maximal piezoelectric properties in the solid solution $Al_{1-x}Sc_xN$ should be exactly equal to $x = 0.43$. But in practice an exact equality could be hardly obtained. In the aim of our experiment, the content of gas mixture N_2/Ar as well as other synthesis parameters was chosen to obtain the $Al_{0.8}Sc_{0.2}N$ composition. One can see by the X-ray diffraction pattern for the test sample ASN/Mo/glass (**Figure 8**) that the ASN film has the preferred orientation (00·2) and the full width at half maximum for this reflection is 0.24°. This shows a good quality texture of axial type along the sixfold axis of the wurtzite-type structure. The X-ray diffraction measurements were performed by Empyrean PANalytical equipment.

A preferred choice of Mo as a bottom electrode was explained by a good accordance between acoustical impedances of diamond substrate and Mo. Aluminum chosen as a material of top

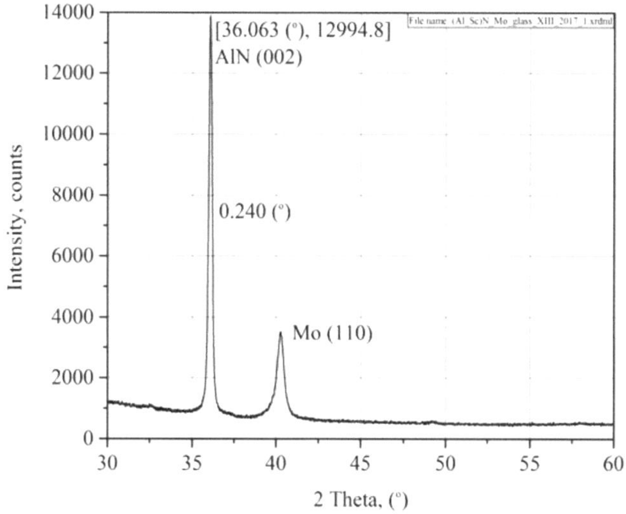

Figure 8. The X-ray diffraction pattern for the test sample ASN/Mo/glass.

electrode can be described as the metal with good conductance and low density. The last property is useful because the deposition of top electrode should be influenced on PLS properties as less as possible. Electrode structures with a specified topology were deposited using photolithography process by the Heidelberg μPG 101 equipment. Explosive photolithography process was necessary to form a specified AlN or ASN film topology. Increasing an accuracy of IDT manufacturing and other electrode structures was connected with an understanding of the physical and chemical features of photolithography process in application to the small-sized substrates. Since the deposition of Al electrodes in the PLS investigated was performed on the surface of AlN or ASN piezoelectric films, it should be taken into account that their surfaces had a distinctive relief defined by triangular pyramid tops of crystallites, so that the roughness of such surfaces can be about 20–30 nm. Deposited Al films should overlap these irregularities. The thickness of deposed metal films was varied within 150–200 nm for the top electrode and 150–200 nm for the bottom one.

5. Microwave investigation tools and measurement methods

Method of low-temperature (LT) microwave studies of PLS acoustic properties was developed on the vector network analyzer E5071C-2 K5 (300 kHz–20 GHz), the probe station M150, the automated low-temperature system for measuring material properties Quantum Design Physical Property Measurement System EverCool II (4–400 K), and the nonstandard LT adapter (**Figure 9**).

To carry out the microwave measurements with HBARs as experimental samples, due to a weak level of the useful signal against the noise, and to obtain the correct quantitative value of the impedance, it must calibrate the whole measuring circuit consisting of a probe, a microwave cable, and a vector network analyzer. Typically, this procedure is performed near the

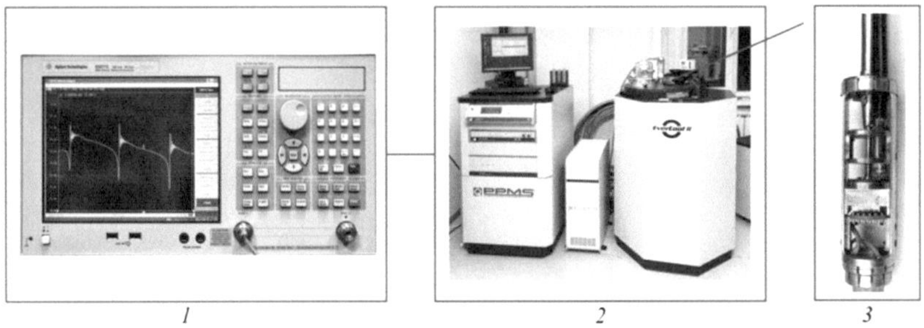

Figure 9. Block diagram of low-temperature microwave measurements: (*1*) vector network analyzer E5071C, (*2*) cryogenic control and temperature stabilization system Quantum Design PPMS EverCool II, and (3) LT measuring adapter.

room temperature at the probe station using a special calibration plate. However, for cryogenic temperatures, the use of a probe station was impossible. We have developed the LT measuring adapter that made possible to measure the complex reflection coefficient $\widehat{S}_{11}(f)$ of microwave signal obtained by HBAR conjugated with the possibility of calibration procedure starting from cryogenic temperatures up to 400 K.

Since all measurements were carried out using a reflected signal, a single-port connection scheme with open-short-load calibration options was selected for measurement and calibration. The calibration elements corresponding to these options were located on the operational disk made of corundum ceramics with contact pads from Au/Pt/Ti produced by the photolithography method. Connection of a measured sample to the contact pads on the operational disk was done by adhering with a silver paste SPI 5001-AB Silver Paint, resistant to both low and high temperatures. At the same time, two measured samples could be placed on the operational disk. Control measurements on the same sample, carried out at room temperature by M150 probe station or using a low-temperature measuring adapter, were in a good agreement.

The method for measuring the sound phase velocity in a substrate was based on the determination of the HBAR's eigenfrequencies taking into account the frequency and temperature dependences of the complex reflection coefficient $\widehat{S}_{11}(f)$ in the composite resonator. Then, the complex impedance $\widehat{Z}_{11}(f)$ of the measuring circuit along with the sample was calculated with the help of vector analyzer software in accordance with the relation:

$$\widehat{Z}_{11}(f) = Z_0 \frac{1+\widehat{S}_{11}(f)}{1-\widehat{S}_{11}}(f), \tag{14}$$

where Z_0 is the input impedance of the microwave circuit equal to 50 Ω. The full impedance $\widehat{Z}_{11}(f)$ included the quantity $\widehat{Z}_{11t}$ of the microwave cable together with the contacts and HBAR's "pure" impedance $\widehat{Z}_{11e}$ as

$$\widehat{Z}_{11}(f) = \widehat{Z}_{11t}(f) + \widehat{Z}_{11e}(f). \tag{15}$$

Required impedance $\widehat{Z}_{11e}(f)$ was determined by Eq. (15) where the value $\widehat{Z}_{11t}$ was experimentally determined at the frequency outside the resonance one of the given overtones. Taking the

$\widehat{Z}_{11e}(f)$ values by the vector analyzer software, the frequency dependence of HBAR's loaded quality factor $Q(f)$ was calculated at a $-$ 3 dB level about the maximum of the Re$\widehat{Z}_{11e}$ module, as well as the values of Re$\widehat{Z}_{11e}$, Im$\widehat{Z}_{11e}$, the phase angle $\varphi(f)$, the group delay time $\tau(f)$, and the Smith diagram. All these parameters quite fully allowed to evaluate HBAR's resonant processes both in a qualitatively and quantitatively sense in the frequency range from 0.6 up to 20 GHz together with a temperature change within 4–400 K.

To determine an absolute value of the BAW phase velocity in the substrate, two different methods were used. Common to both methods was that the antiresonance frequencies $f_{a,n}$, determined about the maxima of Re$\widehat{Z}_{11e}$ module, were measured in the vicinity of an odd half-wave resonance of the thin-film piezoelectric transducer, for example, designated as $f_{p,k}$. The reason was that the near such frequency, within the substrate thickness approximately an integer $(n-1)$ of half-waves, should be fitted at HBAR's excitation on antiresonance points $f_{a,n}$. In this case, the HBAR's quality factor tends to a maximum value. The phase velocity in the substrate V in accordance with the first approach can be calculated by the formula

$$V = \frac{2df_{a,n}}{n-k}, \tag{16}$$

where d is the substrate's thickness and $k = 1, 3, 5, \dots$ is the number of the TFPT half-wave resonance. But an accuracy of the determination of the phase velocity was strongly depended on the correct definition of an overtone number n in the vicinity of the frequency $f_{p,k}$. In the real experiment, it was far from always possible to track all antiresonances, especially in the low-frequency band and bands with high ultrasonic attenuation. In the HBARs investigated, the value of n could reach the values of 200–300 or more, which also made difficult an accurate determination of n.

The second method determining the phase velocity in a substrate was based on the relation obtained taking into account the results (Ref. [17]) about the relation for $\Delta f(N)$ of the upper limit of the spacing between parallel resonant frequencies (SPRF):

$$V \cong 2d\Delta f(N)\left(1 + \frac{\rho_{\mathrm{AlN}} h_{\mathrm{AlN}} + \rho_{\mathrm{Al}} h_{\mathrm{Al}} + \rho_{\mathrm{Mo}} h_{\mathrm{Mo}}}{\rho_{\mathrm{Diam}}}\right), \tag{17}$$

where the corresponding density and thickness of layers or substrate should be inserted in accordance with the order in the "Al/(001) AlN/Mo/(100) diamond" PLS (see **Figure 7a**). In practice, a quantity $\Delta f(N) = f_{a,n+1} - f_{a,n}$ should be chosen in the frequency band called N-region where the maximal Q_n magnitude for nth acoustic overtone of HBAR was measured.

6. Microwave acoustic properties of diamond-based HBARs and SAW resonators utilizing aluminum nitride and aluminum-scandium nitride piezoelectric films

Detailed study of microwave acoustic properties of diamond-based HBARs realized by aluminum nitride and aluminum-scandium nitride piezoelectric films has been fulfilled by a set of

Sample	Piezoelectric layered structure	Thickness (in microns) of			
		Diamond substrate	Piezoelectric film	Top electrode	Bottom electrode
1	Al/(001) AlN/Mo/(100) diamond	299	1.04	0.200	0.150
2	Al/(001) ASN/Mo/(100) diamond	488	1.125	0.140	0.160
3	Al/(001) ASN/Mo/(100) diamond	501	1.125	0.140	0.160
4	Al/(001) AlN/Mo/(100) diamond	180	2.790	0.107	0.135
5	Al/(001) AlN/Mo/(111) diamond	497	2.790	0.107	0.135
6	Al/(001) AlN/Mo/(110) diamond	1274	2.790	0.107	0.135
7	Al/(001) AlN/Mo/(100) diamond	482	0.970	0.105	0.176
8	Al/(001) AlN/Mo/(100) diamond	1107	0.970	0.105	0.176
9	Al IDT/(001) AlN/(001)[110] diamond SAW resonator, d = 10 microns	516	2.790	0.114	—

Table 1. DUT samples based on AlN and ASN piezoelectric layered structures.

Figure 10. The view of the HBAR's sample 3 based on PLS "Al/(001) ASN/Mo/(100) diamond".

samples 1, 2, and 3 (**Table 1**). For example, in **Figure 10** one can see the view of the HBAR's sample 3 based on PLS "Al/(001) ASN/Mo/(100) diamond." On the diamond substrate, two independent HBARs designated as 1a and 1b and differing the effective resonant areas as 20,000 and 5000 square microns, respectively, are located. Topology of the top and bottom electrodes was especially developed to be convenient for an investigation of temperature dependences of HBAR's acoustic parameters within a wide range from 4 up to 400 K.

Results on microwave quality factor Q and effective electromechanical coupling coefficient (EMCC) of diamond-based HBARs (samples 1 and 3) are presented in **Figures 11** and **12**. As one can see from **Figures 11a** or **12a**, the frequency dependence of HBAR's quality factor Q

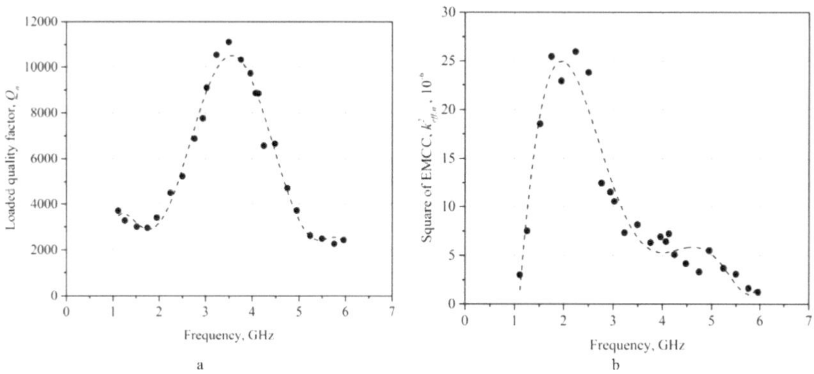

Figure 11. Frequency dependencies of quality factor (a) and the square of EMCC (b) obtained for sample 1 (PLS Al/(001) AlN/Mo/(100) diamond).

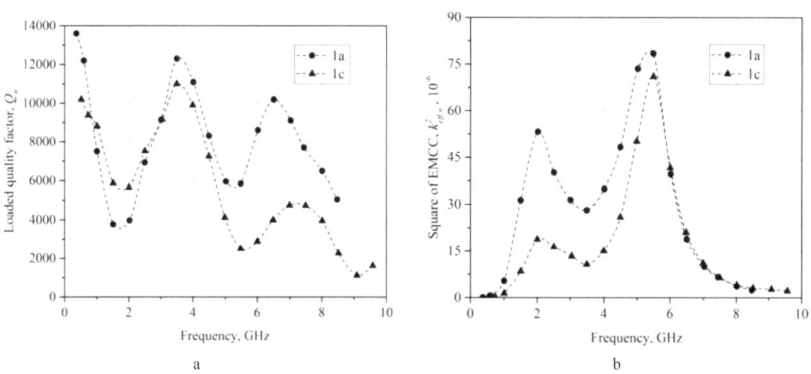

Figure 12. Frequency dependencies of quality factor (a) and the square of the EMCC (b) obtained for sample 3 (PLS Al/(001) ASN/Mo/(100) diamond).

demonstrates a typical behavior observed in piezoelectric layered structures, and such peculiarities as maximal and minimal Q magnitudes at the several frequency bands can be explained by the coinciding of an operational frequency and TFPT eigenfrequencies. Briefly speaking, maximal and minimal Q magnitudes are obtained when the quantities $k\lambda/2$ or $k\lambda/4$ are approximately equal to the TFPT thickness, respectively, and k = 1, 3, 5, …, and λ is the acoustic wave length. For more details see the paper [18].

In order to calculate the square of EMCC, $k^2_{\text{eff},n}$, at a given frequency $\omega_{p,n}$ of parallel resonance of nth acoustic overtone, first the loss resistance R_n was calculated from experimentally obtained impedance Z_{11e} by the relation $R_n = \text{Re}Z_{11e}$. Taking into account the Q_n factor measured for nth acoustic overtone, one can obtain the square of EMCC by the equation taken from Ref. [6]:

$$k^2_{\text{eff},n} = \frac{\omega_{p,n} C_0 R_n}{Q_n}, \tag{18}$$

where C_0 is the TFPT static capacitance.

Then, close to the frequency of the first half-wave resonance of TFPT called N-region according to [16], the value of $n(N)$ for an acoustic overtone with maximal Q_n factor was determined, and the calculation of $k^2_{\text{eff},\,n}(N)$ was carried out using the formula.

$$k^2_{\text{eff},\,n}(N) = \frac{\omega_{p,\,n(N)} C_0 R_{n(N)}}{Q_{n(N)}}. \tag{19}$$

It is interesting to compare the AlN and ASN piezoelectric properties taking into account such electromechanical coupling coefficient as the k_t quantity which is defined for the thickness-extensional mode of a conventional piezoelectric resonator and can be calculated taking into account the piezoelectric, elastic, and dielectric properties of a piezoelectric material:

$$k_t^2 = \frac{e_{33}}{C^E_{33}\varepsilon^S_{33}}, \tag{20}$$

where e_{33}, C^E_{33}, and ε^S_{33} are the piezoelectric, dielectric, and elastic constants, respectively. The second and last ones should be measured at the constant electric field and at the constant strain conditions, respectively.

Following to [16], the relation between k_t^2 and $k^2_{\text{eff},\,n}(N)$ can be written in the form.

$$k_t^2(N) \cong \frac{\rho_{\text{AlN}} h_{\text{AlN}} (\rho_{\text{AlN}} h_{\text{AlN}} + \rho_{\text{Al}} h_{\text{Al}} + \rho_{\text{Mo}} h_{\text{Mo}} + \rho_{\text{Diam}} h_{\text{Diam}})}{(\rho_{\text{AlN}} h_{\text{AlN}} + \rho_{\text{Al}} h_{\text{Al}} + 0.5\rho_{\text{Mo}} h_{\text{Mo}})^2} k^2_{eff,\,n}(N) \tag{21}$$

where the corresponding density and thickness of layers or substrate are the same as in Eq. (21). Data on k_t^2 values obtained for the samples 1–3 were summarized in **Table 2**. For of true comparison of results, the samples with the close dimensions and shape of TFPT have to be chosen. So, the samples 1 and 2 based on ASN piezoelectric film were obtained in the same process. Then, the overtones with close resonant frequencies have been selected. Note that studied HBARs based on PLSs differing the material of piezoelectric films and substrate thickness demonstrate the close magnitudes of quality factor Q ~ 11,000–12,000 which corresponds to comparatively high-quality parameter $Q{\cdot}f \sim 4.2{\cdot}10^{13}$ Hz at 3500 MHz.

As a main result, one can emphasize that the application of aluminum-scandium nitride as piezoelectric material leads to a drastic increase of both the effective $k^2_{\text{eff},\,n}$ and k_t^2 EMCCs up to 2.5 times. Other things being equal, the ASN-based acoustoelectronic devices will have the prospective advantages.

Sample	PLS	Frequency (MHz)	$k^2_{\text{eff},\,n}(N)$, 10^{-5}	k_t^2, 10^{-3}	k_t (%)
1	Al/(001) AlN/Mo/(100) diamond	3550	0.82	1.34	3.7
2	Al/(001) ASN/Mo/(100) diamond	3500	3.0	8.85	9.4
3	Al/(001) ASN/Mo/(100) diamond	3500	2.81	7.82	8.8

Table 2. Data on EMCCs of the HBAR samples investigated.

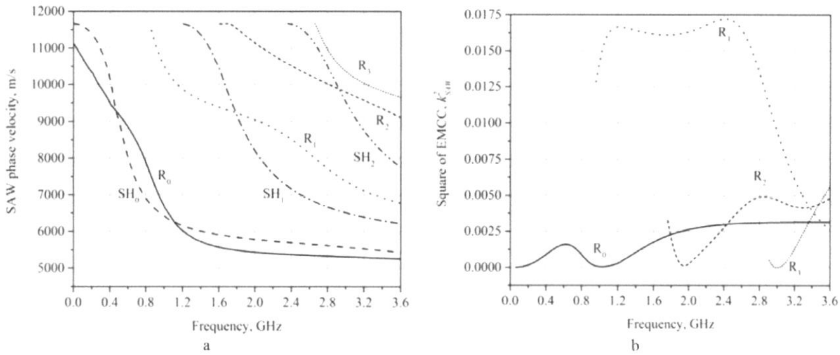

Figure 13. Frequency dependencies of SAW phase velocities (a) and the square of the EMCC (b) calculated for sample 9 (PLS "Al IDT/(001) AlN/(001)[110] diamond"). Curves designated as R_0, R_1, R_2, … and SH_0, SH_1, SH_2, … should be associated with Rayleigh or surface horizontal-type waves, respectively.

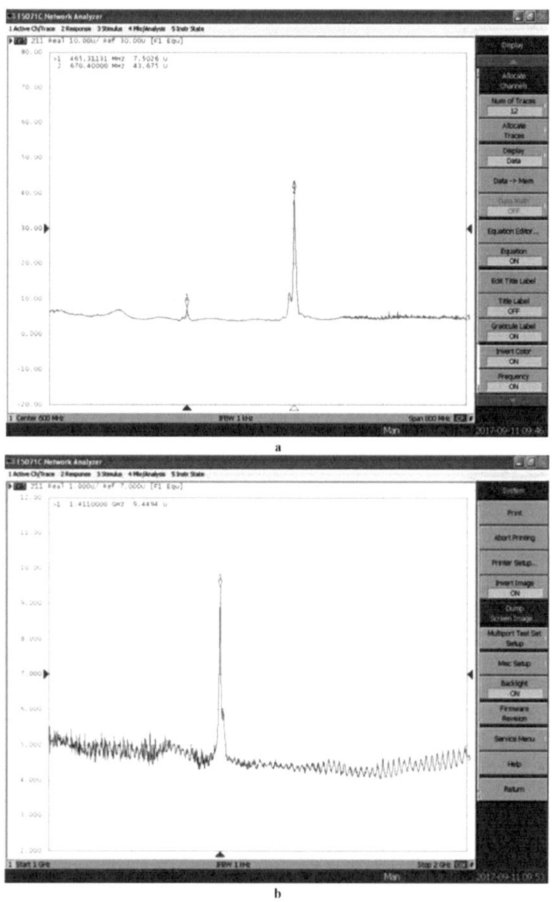

Figure 14. Amplitude and frequency characteristics of DUT sample 9 based on PLS "Al IDT/(001) AlN/(001)[110] diamond" at room temperature: (a) impedance Z_{11} in the frequency band 400–600 MHz and (b) impedance Z_{11} in the vicinity of 1400 MHz. The number of MSR electrode pairs was equal to $N = 32$.

As a DUT sample 9 (see **Table 1**), the SAW resonator based on PLS "Al IDT/(001) AlN/(001) [110] diamond" was investigated in the frequency band from 400 up to 1500 MHz at the SAW propagation in the [110] direction on the (001) crystalline cut of diamond. Scheme of a single-port SAW resonator was presented in **Figure 7b**. Distance between the nearest electrodes of IDT and MSR was taken as d = 10 microns at the width of an electrode as 5 microns. The number of MSR electrode pairs was varied from N = 16 up to 128. Because the SAW propagation was studied in the piezoelectric layered structure, the last one should be considered as a thin-film waveguide. In this case the dispersion of SAW phase velocities as well as other parameters must be taken into account. As an example such dispersion dependences of SAW phase velocities and EMCC were calculated for the PLS considered (**Figure 13**).

The square of the EMCC concerned with surface acoustic waves was calculated by a conventional relation as.

$$k_{\mathrm{SAW}}^2 = 2\frac{V_{\mathrm{SAW}} - V_{\mathrm{SAW},s}}{V_{\mathrm{SAW}}}, \tag{22}$$

where V_{SAW} and $V_{\mathrm{SAW},s}$ should be defined at a free or shorted surface boundary conditions. In the last one, a conductive layer with an infinitesimal thickness should cover the surface of a piezoelectric crystal. Analyzing **Figure 13b**, one can conclude that the waves of surface horizontal type cannot be excited at a given orientation of piezoelectric film. But the Rayleigh-type waves could be excited at some frequency bands distinctive for waves of the different orders. Note that the EMCC of the R_1 mode, so-called Sezawa wave, should have a considerably more value than that in conventional Rayleigh mode R_0. Experimental data obtained by sample 9 are presented in **Figure 14**. Resonance frequencies of the SAW modes were detected as 465, 672, and 1412 MHz. Taking into account the data in **Figure 14**, one can define the type of SAW

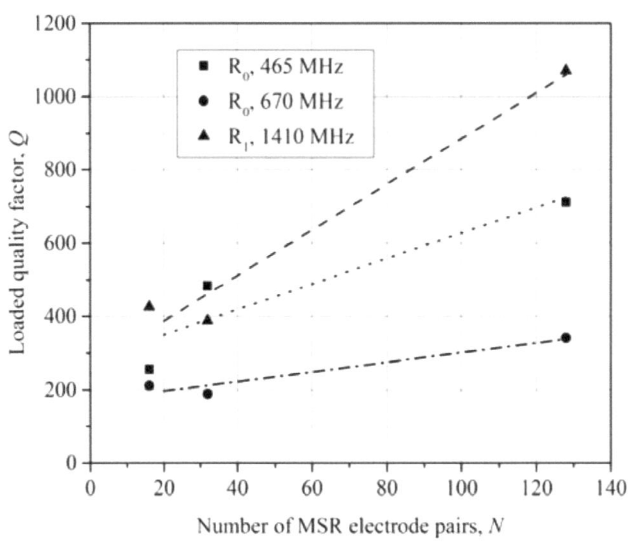

Figure 15. Dependence of the quality factor of SAW resonators in the of DUT sample 9 on the number of MSR electrode pairs for three SAW modes (PLS "Al IDT/(001) AlN/(001)[110] diamond").

modes as the first (465 MHz) and second (672 MHz) harmonics of R_0 mode and the third harmonics of R_1 mode (1412 MHz). It was obtained that the quality factor of SAW resonators was strongly depended on the number of MSR electrode pairs (**Figure 15**). This corresponds to lower-energy losses in SAW resonator and a decrease in the passband of the frequency response. The highest $Q \approx 1050$ was observed for the R_1 Sezawa mode at the resonant frequency 1412 MHz.

7. Investigation of temperature dependences of acoustic parameters in diamond-based piezoelectric layered structures

In a way described in **Section 5**, temperature dependences of the loaded quality factor Q_n for a lot of acoustic overtones and the phase velocity V of longitudinal acoustic wave in the [100] direction of synthetic IIa type diamond are shown in **Figure 16**. As a DUT the HBAR sample 4 based on the piezoelectric structure "Al/AlN/Mo/(100) diamond" (see **Table 1**) was chosen. For this sample, the measurements were performed at the operating frequency of 5.41 GHz. Value of the phase velocity V = 17,542 m/s obtained at room temperature coincides with that measured earlier using the pulse-phase method operating on the frequencies of 10–200 MHz [18]. When calculating the temperature dependence of both the phase velocity and elastic modulus, the change in the thickness and density of a diamond substrate due to linear thermal expansion was taken into account by the data in Ref. [19]. In that case, the temperature dependence of an elastic modulus C_{11} of diamond was calculated in accordance with the relation.

$$C_{11}(T) = \rho(T)[V(T)]^2, \tag{23}$$

where in those calculations the diamond's density at room temperature ρ= 3516 kg/m^3 was taken, and $V(T)$ was the calculating value of the phase velocity of longitudinal acoustic wave propagating along the in [100] crystalline direction of diamond at a given temperature. First-order temperature coefficient of the BAW velocity was evaluated as $5.1{\cdot}10^{-6}$ K^{-1} in the vicinity of a room temperature. Comparison of our data on the temperature dependence of diamond's

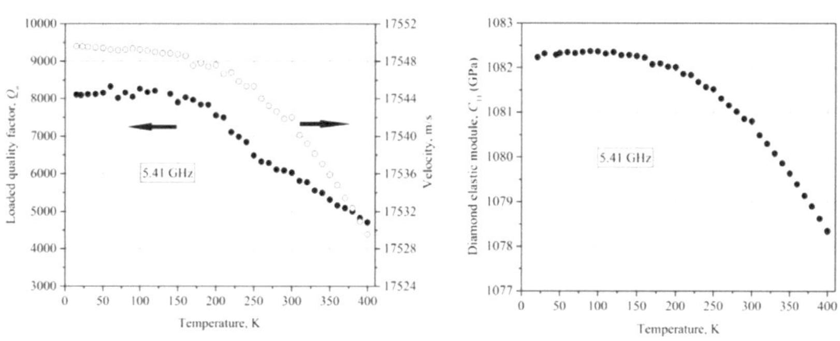

Figure 16. Temperature dependences of a loaded Q-factor and BAW phase velocity in the direction [100] of synthetic diamond (a) and the C_{11} elastic modulus of diamond (b) obtained by sample 4.

elastic modulus C_{11} obtained in the temperature range 10–303 K, with the corresponding one in Ref. [20] measured at an operational frequency of up to 60 MHz in the temperature range 77–323 K, has shown a good qualitative agreement. Some quantitative difference can be, firstly, due to the distinction in the objects of research, since we have used the samples based on the IIa type synthetic diamond, and in Ref. [20] a natural diamond has been investigated, and, secondly, with the fact that the value of diamond density 3516 kg/m^3 in our calculations has been differed from 3512 kg/m^3 which was used in Ref. [20].

Qualitatively, the temperature changes in the elastic modulus correspond to the conventional variation of elastic properties in a lot of solid states. One can emphasize on the existence of two temperature regions at which a clearly distinctive behavior of elastic properties as well as quality factor was observed: firstly, a weak dependence in the low-temperature region, which corresponds to the "freezing out" of the acoustic phonons and, secondly, a well-formed dependence close to linearly proportional one in the relatively high-temperature region, where, according to Planck's law, a lot of acoustic phonons will be excited (**Figure 16)**. Note that the temperature dependences of the elastic moduli at higher temperatures are related to the lattice anharmonicity: if the temperature is raised, the effective lattice stiffness decreases by increasing the distance between ions participating in the lattice vibrations. This ultimately leads to a nonlinear interaction between the phonons.

Research of the temperature dependence of HBAR parameters was performed with using a set of samples which were different in crystallographic orientations of diamond substrate, a thickness of films of aluminum nitride and aluminum-scandium nitride, an electrode topology, etc. The choice of a given piezoelectric film thickness was mainly associated with the TFPT peculiarities: if it is desirable to get a more TFPT effectiveness at low frequencies, it should necessarily deposit a higher thickness. So, in a way to obtain the results at the frequency band below 1 GHz, the appropriate TFPTs were based on AlN films with the thickness up to 2.79 microns (see **Table 1**, samples 4–6). Besides it, the choice of an operational frequency should be agreed with the frequency bands where the maximal Q-factor will be observed (see **Section 6**). All the results (**Figure 17**) on the temperature dependences of a loaded quality factor $Q(T, f)$ in a region from 10 up to 400 K at different frequencies from 0.8 up to 4.9 GHz were obtained by the samples 3–8 based on the (100), (110), and (111) diamond substrates (see **Table 1)**.

In **Figure 17a** the normalized magnitudes of Q-factor have been presented when the reference $Q_{400\ K}$ values were taken at the point 400 K, and the absolute Q values were displayed in **Figure 17b** and **c**.

It should be emphasized that an evident general conclusion, such as the results obtained at relatively low frequencies below 1.1 GHz, was quite different in comparison with ones measured at the frequencies up to 5 GHz. Such behavior was observed for all the samples regardless the substrate's orientation. Really, observing the Q-factor variation under the temperature decrease, one can see a considerable increasing Q value with further achievement of a maximal magnitude at low temperatures, when the high-frequency acoustic overtones were taken into account. On the other hand, the Q-factor of overtones excited on the frequencies below 1.1 GHz had only a monotonic temperature dependence as a slightly Q increasing at low temperatures. We assume that the difference could be explained in terms of changing

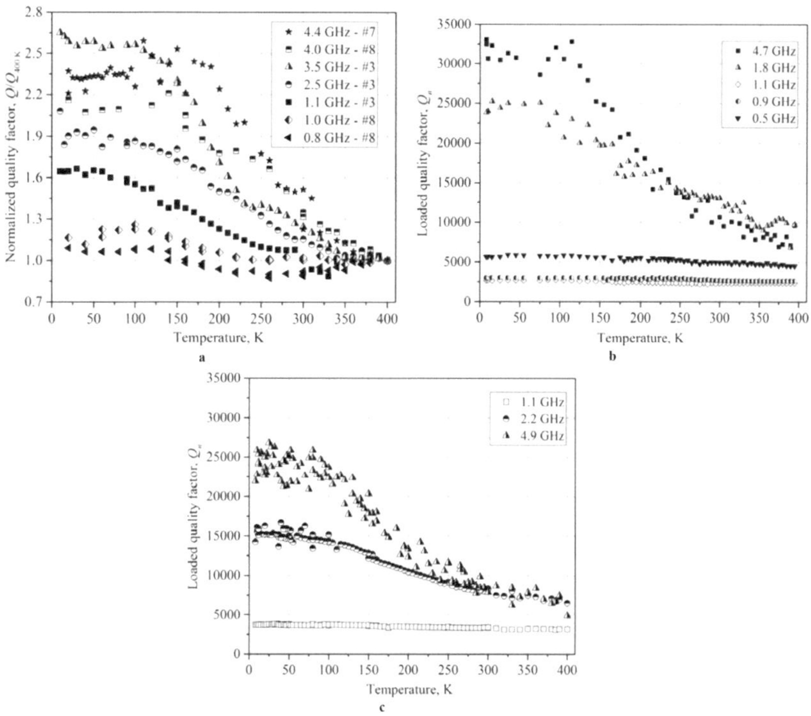

Figure 17. Temperature and frequency dependences of a normalized loaded Q-factor in the HBAR's samples 3, 7, and 8 with the propagation of microwave longitudinal waves along the [100] diamond's direction (a); temperature and frequency dependences of the quality factor Q_n in the HBAR's sample 6 with the propagation of microwave longitudinal waves along the [110] diamond's crystalline direction (b); temperature and frequency dependences of the quality factor Q_n in the HBAR's sample 5 with the propagation of microwave longitudinal waves along the [111] diamond's crystalline direction (c).

mechanism of acoustic attenuation from Akhiezer's type at the frequencies below 1.1 GHz to Landau-Rumer's one at a higher frequency band in the single-crystalline diamond. In the last one, if taking a given point close to room temperature, the so-called quality parameter $Q{\cdot}f$ with the frequency variation must change as a linearly proportional function in a manner of $Q{\cdot}f \sim f$, i.e., the quality factor must not depend on the frequency (Ref. [21]) in that approximation. Earlier, we have obtained by means of HBAR based on PLS involved the single-crystalline diamond substrate and AlN thin-film piezoelectric transducer that such assumption has been quite fulfilled for operational frequencies up to 20 GHz at room temperature (Ref. [7]). Also, this fact is of great practical importance, as it allows realizing acoustoelectronic devices on diamond substrates at hypersonic frequencies with an acceptable level of acoustic attenuation.

8. Conclusion

Theory of the influence of an external loading on the acoustic parameters of piezoelectric five-layered structure as "Al/(001) AlN/Mo/(001) diamond/Me" has been derived. Approach how to obtain the boundary conditions can be spread on the more complicated multilayered structures too. On the base of that theory, the own software "Modeling of the processes of resonant acoustic spectroscopy in multilayered structures" has been developed. Study of the influence of metal film deposition as the fifth layer on the change of the overtone resonant frequency of diamond-based PLS "Al/(001) AlN/Mo/(001) diamond/Me" (Me = Al, Mo) has been fulfilled. It has been obtained that metal film deposition leads to the decrease of resonant frequency with the thickness increasing both the metals.

HBAR based on the PLS "Al/(001) AlN/Mo/(001) diamond" has been investigated in terms of 3D FEM simulation by the software COMSOL Multiphysics. Boundary conditions on the top and bottom surfaces of a model sample were chosen as the free ones, and on all the vertical surfaces as the symmetrical ones. All the dimensions of a model sample were close to appropriate ones in the experimental sample. Besides Lamb waves the BAW and SAW modes of Rayleigh and SH type were found. It has been shown that the calculated data on HBAR's resonant frequency in 2D approximation were in close accordance with the similar obtained in 3D approximation. Elastic displacement patterns arising for the SH-type waves in the PLS "Al/(001) AlN/Mo/(001) diamond" were obtained in 3D image. Dispersive dependences of SH-type wave velocities on a frequency for a number of the first dispersive SH-branches have been calculated. Comparison of the data obtained by 3D FEM simulation with the theory and 2D FEM results has been demonstrated a good agreement between a model and theory. Visualization of elastic displacement fields associated with the SAW of Rayleigh type has been realized, and calculated phase velocity of the SAW propagating in a given direction on the (001) diamond surface was quite the same as known data.

Peculiarities of the technology of aluminum nitride and aluminum-scandium nitride piezoelectric films have been discussed. The $Al_{0.8}Sc_{0.2}N$ composition was obtained to create the microwave BAW and SAW test devices as diamond-based HBAR and SAW resonator. By the X-ray diffraction pattern, it has been proven that the crystallites of ASN film had a preferred orientation (00·2) and the full width at half maximum for that reflection was 0.24°. This shows on a good quality texture of axial type along the sixfold axis of the wurtzite-type structure. Data obtained will be used for the future development of ASN film technology in a way of producing a number of compositions with the better piezoelectric properties.

Detailed study of microwave acoustic properties of diamond-based HBARs realized by aluminum nitride and aluminum-scandium nitride piezoelectric films has been fulfilled by a set of the samples. Topology of the top and bottom electrodes as well as piezoelectric film areas was especially developed to be convenient for an investigation of temperature dependences of HBAR's acoustic parameters within a wide range from 4 up to 400 K. Investigated HBARs based on PLSs differing the material of piezoelectric films and substrate thickness have demonstrated the close magnitudes of quality factor $Q \sim 11{,}000–12{,}000$ which corresponded to comparatively high-quality parameter $Q{\cdot}f \sim 4.2{\cdot}10^{13}$ Hz at 3500 MHz. As a main result, one can emphasize that the application of aluminum-scandium nitride as a piezoelectric material has resulted in a drastic increase of the effective electromechanical coupling coefficient up to

2.5 times in comparison with that in aluminum nitride. Other things being equal, the ASN-based acoustoelectronic devices will have the prospective advantages.

Single-port SAW resonator based on PLS "Al IDT/(001) AlN/(001)[110] diamond" has been investigated in the frequency band from 400 up to 1500 MHz at the SAW propagation in the [110] direction on the (001) crystalline cut of diamond. Highest $Q \approx 1050$ was observed for the Sezawa mode at the resonant frequency 1412 MHz. Note that the microwave SAW devices have demonstrated a comparatively low performance in comparison with diamond-based HBAR taken at the same operational frequency.

Method of HBAR microwave studies of temperature dependences of such acoustic properties as BAW phase velocity and quality factor in the temperature region 4–400 K and frequency band 0.5–5 GHz has been developed. A general conclusion should be emphasized that the results on the temperature dependence of diamond's Q-factor obtained at relatively low frequencies below 1.1 GHz were quite different in comparison with the ones measured at the frequencies up to 5 GHz. Such behavior was observed for all the samples regardless the substrate's orientation. One can assume that the difference could be explained in terms of changing mechanism of acoustic attenuation from Akhiezer's type at the frequencies below 1.1 GHz to Landau-Rumer's one at a higher frequency band in the single-crystalline diamond. In the last case, if taking a given point close to room temperature, the quality parameter $Q{\cdot}f$ with the frequency variation must change as a linearly proportional function in a manner of $Q{\cdot}f \sim f$, i.e., the quality factor must not depend on the frequency in that approximation. That result has earlier been proven for operational frequencies up to 20 GHz at room temperature [7]. Also, this fact is of great practical importance, as it allows realizing acoustoelectronic devices on diamond substrates at hypersonic frequencies with an acceptable level of acoustic attenuation.

Acknowledgements

This work was supported by a grant of the Russian Science Foundation (project 16-12-10293).

Author details

Boris P. Sorokin[1,2]*, Gennady M. Kvashnin[1], Andrey S. Novoselov[1,2], Sergey I. Burkov[3], Anton B. Shipilov[1,2], Nikolay V. Luparev[1], Victor V. Aksenenkov[1] and Vladimir D. Blank[1]

*Address all correspondence to: bpsorokin1953@yandex.ru

1 Technological Institute for Superhard and Novel Carbon Materials, Troitsk, Moscow, Russian Federation

2 Moscow Institute of Physics and Technology, Dolgoprudny, Moscow Region, Russian Federation

3 Siberian Federal University, Krasnoyarsk, Russian Federation

References

[1] Lakin KM. Thin film resonator technology. IEEE Transactions on Ultrasonics, Ferroelectrics, and Frequency Control. 2005;**52**:707-716. DOI: 10.1109/TUFFC.2005.1503959

[2] Part 4. Acoustic wave based microdevices. In: Marco G Beghi, editor. Acoustic Waves—From Microdevices to Helioseismology. Rijeka, Croatia: Intech; 2011. pp. 419-652. DOI: 10.5772/1032

[3] Akiyama M, Kamohara T, Kano K, Teshigahara A, Takeuchi Y, Kawahara N. Enhancement of piezoelectric response in scandium aluminum nitride alloy thin films prepared by dual reactive cosputtering. Advanced Materials. 2009;**21**:593-596. DOI: 10.1002/adma.200802611

[4] Zhang H, Pang W, Yu H, Kim ES. High-tone bulk acoustic resonators on sapphire, crystal quartz, fused silica, and silicon substrates. Journal of Applied Physics. 2006;**99**:124911. DOI: 10.1063/1.2209029

[5] Baumgartel L, Kim ES. Experimental optimization of electrodes for high *Q*, high frequency HBAR. In: Proceedings of 2009 IEEE Intl Ultrasonics Symp. Rome, Italy; 2009. pp. 2107-2110. DOI: 10.1109/ULTSYM.2009.5441814

[6] Mansfel'd GD, Alekseev SG, Polzikova NI. Unique properties of HBAR characteristics. In: Proceedings of the 2008 IEEE Intl Ultrasonics Symp. Beijing, China; 2008. pp. 439-442. DOI: 10.1109/ULTSYM.2008.0107

[7] Sorokin BP, Kvashnin GM, Novoselov AS, Bormashov VS, Golovanov AV, Burkov SI, Blank VD. Excitation of hypersonic acoustic waves in diamond-based piezoelectric layered structure on the microwave frequencies up to 20 GHz. Ultrasonics. 2017;**78**:162-165. DOI: 10.1016/j.ultras.2017.01.014

[8] Benetti M, Cannata D, Fabio Di Pietrantonio FD, Verona E. Growth of AlN piezoelectric film on diamond for high-frequency surface acoustic wave devices. IEEE Transactions on Ultrasonics, Ferroelectrics, and Frequency Control. 2005;**52**:1806-1811. DOI: 10.1109/TUFFC.2005.1561635

[9] Fujii S, Kawano S, Umeda T, Fujioka M, Yoda M. Development of a 6 GHz resonator by using an AlN Diamond structure. In: Proceedings of the 2008 IEEE Intl Ultrasonics Symp. Beijing, China; 2008. pp. 1916-1919. DOI: 10.1109/ULTSYM.2008.0472

[10] Aleksandrov KS, Sorokin BP, Burkov SI. Effective Piezoelectric Crystals for Acoustoelectronics, Piezotechnics and Sensors. Vol. 1. Novosibirsk: SB RAS Publishing House; 2007. p. 501. (in Russian)

[11] Alekseev SG, Kotelyanskii IM, Polzikova NI, Mansfel'd GD. Study of layered structures using modified acoustic resonator spectroscopy. Journal of Communications Technology, Electronics. 2015;**60**:300-307. DOI: 10.1134/S1064226915030018

[12] Sorokin BP, Burkov SI. Modeling of the processes of resonant acoustic spectroscopy in multilayered structures. Certificate #2017660543 of government registration of computer software. Russian Federation; 2017

[13] Sorokin BP, Kvashnin GM, Telichko AV, Burkov SI, Blank VD. Piezoelectric layered structures based on the synthetic diamond. In: Ogawa T, editor. Piezoelectric Materials. Rijeka, Croatia: Intech; 2016. pp. 161-199. DOI: 10.5772/62630

[14] Sorokin BP, Kvashnin GM, Telichko AV, Novoselov AS, Burkov SI. Lamb waves dispersion curves for diamond based piezoelectric layered structure: Experimental investigation and computer modeling. Applied Physics Letters. 2016;**108**:113501. DOI: 10.1063/1.4943945

[15] Sorokin BP, Bormashov VS, Korostilev EV, Novoselov AS, Doronin MA, Kravchuk KS, Blank VD. Usage of electron back scattering diffraction for investigation of buried damage layer underneath a single crystalline diamond surface. Journal of Materials Science: Materials in Electronics. 2017;**28**:13464-13471. DOI: 10.1007/s10854-017-7185-y

[16] Teshigahara A, Hashimoto K, Akiyama M. Scandium aluminum nitride: Highly piezoelectric thin film for RF SAW devices in multi GHz range. In: Proceedings of 2012 IEEE Intl Ultrasonics Symp. Dresden, Germany; 2012. pp. 1-5. DOI: 10.1109/ULTSYM.2012.0481

[17] Zhang Y, Wang Z, Cheeke JDN. Resonant spectrum method to characterize piezoelectric films in composite resonators. IEEE Transactions on Ultrasonics, Ferroelectrics, and Frequency Control. 2003;**50**:321-333. DOI: 10.1109/TUFFC.2003.1193626

[18] Sorokin BP, Kvashnin GM, Kuznetsov MS, Telichko AV, Burkov SI. Influence of the temperature and uniaxial pressure on the elastic properties of synthetic diamond single crystal. In: Proceedings of 2012 IEEE Intl Ultrasonics Symp. Dresden, Germany; 2012. pp. 763-766. DOI: 10.1109/ULTSYM.2012.0190

[19] Novikov NB, editor. Physical Properties of Diamond (in Russian). Handbook. Naukova dumka: Kyiv; 1987. p. 188

[20] McSkimin HJ, Andreatch PJ. The elastic stiffness moduli of diamond as a function of pressure and temperature. Applied Physics. 1972;**43**:2944-2949. DOI: 10.1063/1.1661636

[21] Tabrizian R, Rais-Zadeh M, Ayazi F. Effect of phonon interactions on limiting the $f{\cdot}Q$ product of micromechanical resonators. In: Proceedings of 15 Intl Conf on Solid-State Structures, Actuators and Microsyst. Denver, USA; 2009. pp. 2131-2134. DOI: 10.1109/SENSOR.2009.5285627

Chapter 3

Hydrodynamic Loading on Vibrating Piezoelectric Microresonators

Huacheng Qiu and Helmut Seidel

Additional information is available at the end of the chapter

http://dx.doi.org/10.5772/intechopen.77731

Abstract

The dynamics of micro-piezoelectric resonators can be profoundly affected by immersion in fluids. Aluminum nitride-based piezoelectric microresonators are fabricated and tested under controlled pressures in several gases. The cases on microresonator vibrating in fluid can be broadly divided into: (i) those that deal with vibration in free space and (ii) close to a surface. For the first case, experimental and analytical results for the hydrodynamic loading characteristics of the resonators at different resonant modes have been investigated, as well as the influences of fluid viscosity and compressibility. For the second case, most prior efforts have been focused on squeeze-film damping with very narrow gaps, while in many practical applications, the resonators vibrate close to a surface with a moderate distance. Experiments by using a micro-bridge resonator with a big range of gaps are performed and compared with predictions from theoretical models.

Keywords: aluminum nitride, fluid-structure interaction, viscous damping, compressibility, wall effect

1. Introduction

Piezoelectric microresonators have been recently used in a large variety of applications [1–9], due to their tiny structures [10] and ultrahigh sensitivity [11], let alone their self-exciting and self-sensing capability, together with full integration, as compared to electrostatic [12], electromagnetic [13], or optically associated [14] structures. Regarding the piezoelectric materials, aluminum nitride (AlN) was selected in this work, to be compatible with complementary metal-oxide-semiconductor (CMOS) technology. Besides, by carefully controlling the fabrication parameters [15], AlN film with high piezoelectric coefficients and low intrinsic stress can be obtained.

Microresonators are normally operated in fluids (such as air or liquid) at atmosphere or under reduced pressure. The resonator parameters, such as resonance frequency f_r, quality factor Q or phase shift, may change due to environmental influences. The Q factor is defined as the ratio of stored energy to the dissipated energy per cycle, equivalents to $f_r/\triangle f$, where $\triangle f$ is the peak width at half power. In general, a high Q factor is preferred for micro-resonator sensors, as high Q factor results in a sharper resonance peak and a better detectable resolution.

The energy loss Q^{-1} for a resonator consists of intrinsic energy loss Q_{int}^{-1} (e.g., radiation of elastic energy into attachment and structural friction) [16] and external losses Q_{flu}^{-1} (e.g., acoustic, viscous, and squeeze-film damping) into the surrounding fluid. Q_{int}^{-1} can be measured by operating the resonator in a high vacuum. When operating in a fluidic environment, Q_{flu}^{-1} would be the major energy loss source. The cases on microresonator vibrating in fluid can be broadly divided into (i) those that deal with vibration in free space and (ii) close to a surface. Experimental and analytical results for the hydrodynamic loading characteristics of the resonators in both cases are presented in this chapter.

2. Experimental techniques

In this section, we describe the experimental techniques, including the device fabrication, packaging, measurement setup, and electric readout. Piezoelectric AlN-based resonators were fabricated and tested.

2.1. AlN-based piezoelectric microresonator fabrication

The AlN thin films were deposited using reactive sputtering. Films with good c-axis orientation [15] have been successfully achieved, and the effective piezoelectric constants were measured as: d_{33} = 3 pm/V and d_{31} = −1.0 pm/V [17].

Low resistivity (less than 0.1 Ω cm) p-type (100) single crystal silicon (SCS) wafer was used as a resonating element, serving simultaneously as a bottom electrode. The fabrication process is shown as in **Figure 1**. The SCS wafer was oxidized in high temperature furnace to form 120 nm thick SiO_2 layers on both sides, and then a 500 nm thick PECVD Si_xN_y layer was deposited on the bottom side of the wafer (**Figure 1a**). In a next step, a 1 μm thick AlN film and a 300 nm thick gold film were deposited on the top side and etched to form a sandwiched piezo-electrode stack (**Figure 1b**). The resonant beam was formed by back side wet etching and then released by a dry etching process (**Figure 1c**). **Figure 2** shows an optical micrograph of a fabricated resonator.

2.2. Device packaging

Two resonator packages were used to investigate the hydrodynamic loading effects in free space or close to a surface. For vibration in free space, the resonator chip was mounted on a printed circuit board, with a pre-drilled hole (c.f. **Figure 3a** and **b**), and the resonator is suspended several millimeters away from any surface, much bigger than the resonator dimensions. For vibration

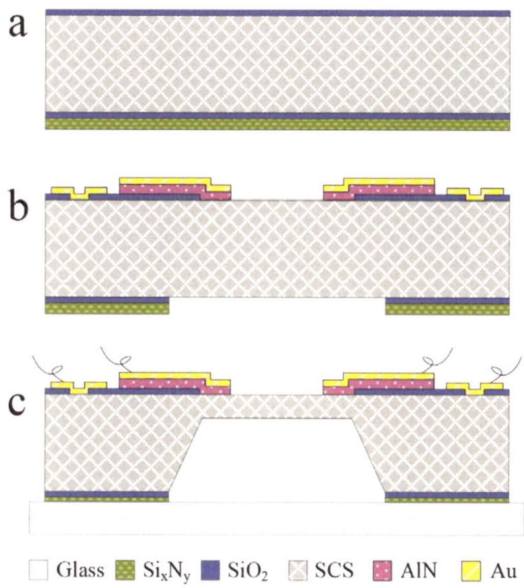

Figure 1. Flow chart of the fabrication process.

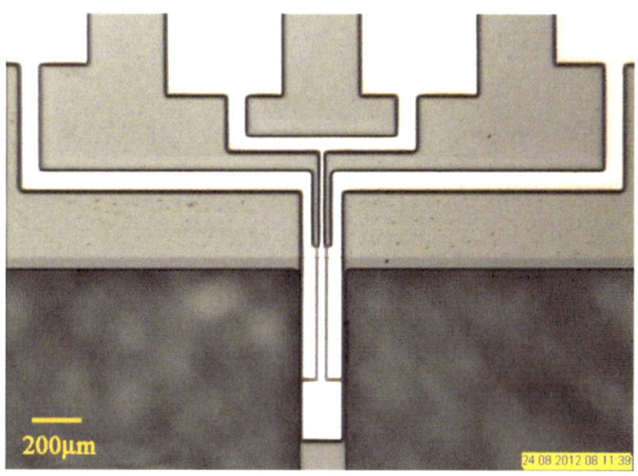

Figure 2. Micrograph of a piezoelectric resonator.

close to a surface, a curved cover was clamped on the top surface of the resonator (c.f. **Figure 3c, d**), and the cavity depth h_0 of the cover varies from 20 to 300 μm.

2.3. Measurement setup

The packaged resonator is then mounted on an electronic circuit (whose major functions are electrical crosstalk compensation and signal amplifier, for details one can refer to [18, 19]) and then placed in a custom-built vacuum chamber, wherein the pressure can be controlled from atmospheric pressure down to high vacuum (lower than 10^{-4} mbar). Meanwhile, five noble gases (He, Ne, Ar, Kr, and Xe) and N_2 are used to observe the resonator performance variation.

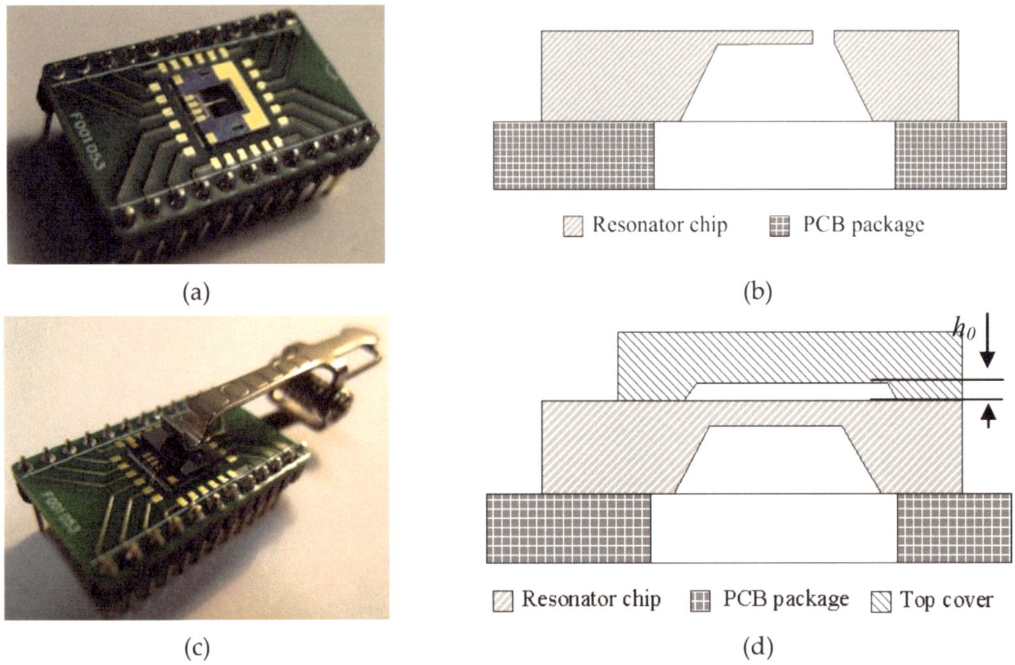

Figure 3. The photos (a and c) and schematic cross-sectional views (b and d) of packaged chips for vibration in free space and close to a surface, respectively.

The noble gases are chosen in this work, since they are all monatomic gases with gradually increasing density and decreasing sound speed under standard conditions. **Figure 4** shows (a) the camera picture (the gas bottles are behind the chamber, not shown in the picture) and (b) the schematic diagram of the measurement setup used to characterize the resonator behavior.

2.4. Electric readout

The resonance frequencies and Q factors of the packaged resonators were measured and characterized under controlled pressures. For vibrations in a high vacuum (10^{-4} mbar), a high Q value can be achieved and the fluidic damping is negligible. Increasing the ambient pressure results in a slight shift in f_r but a dramatic reduction in the Q factor, as shown by example resonant curves in **Figure 5**.

3. Hydrodynamic loading on vibrating microresonators

In this section, the research interest is focused on beam-shaped resonators operating in gas media. First, the full set of Navier-Stokes (N-S) equation is semi-analytically solved, and the solution can be interpreted by a "three wave theory" with coupled viscous, thermal, and acoustic waves. Second, for slender structures at moderate reduced pressures, viscous drag is typically the dominant loss mechanism. The influence of the fluid's viscosity, density, and compressibility on the energy loss of the vibrating beam is discussed. Third, in many practical

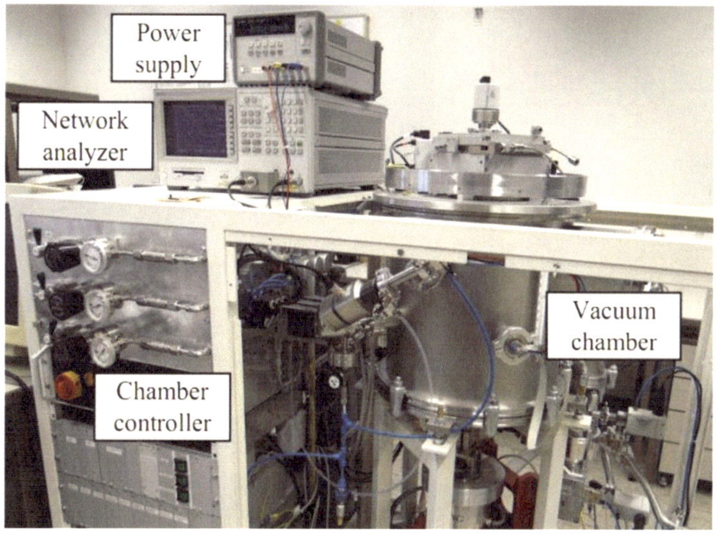

(a)

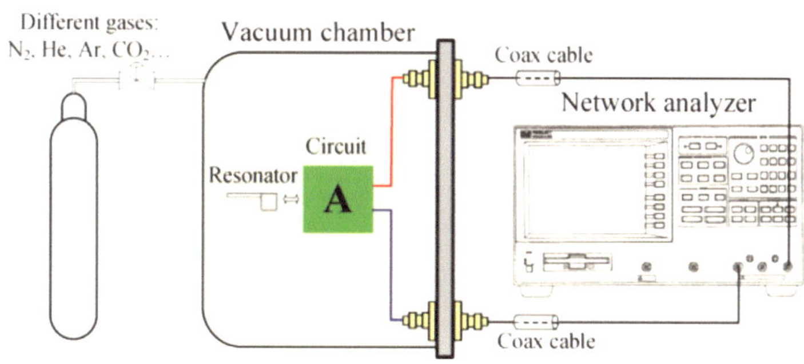

(b)

Figure 4. (a) The camera picture and (b) the schematic diagram of the measurement setup used to characterize the microresonators in the vacuum chamber.

applications, resonators are vibrating near a surface, and the influence of the vicinity on such a surface is experimentally and analytically evaluated.

3.1. Full N-S model and numerical simulation

For a continuous fluid, the most extensive type of theoretical model is based on the full set of N-S equations, which takes into account of fluid compressibility, thermal conductivity, and viscosity. The variables are density, pressure, temperature, and velocity that all vary around the resonator.

3.1.1. Theory

The basic equations governing the fluid motion induced by vibration are the compressible N-S equation, the equation of continuity, the equation of state for an ideal gas, and the energy equation [20]. The equations can be written as:

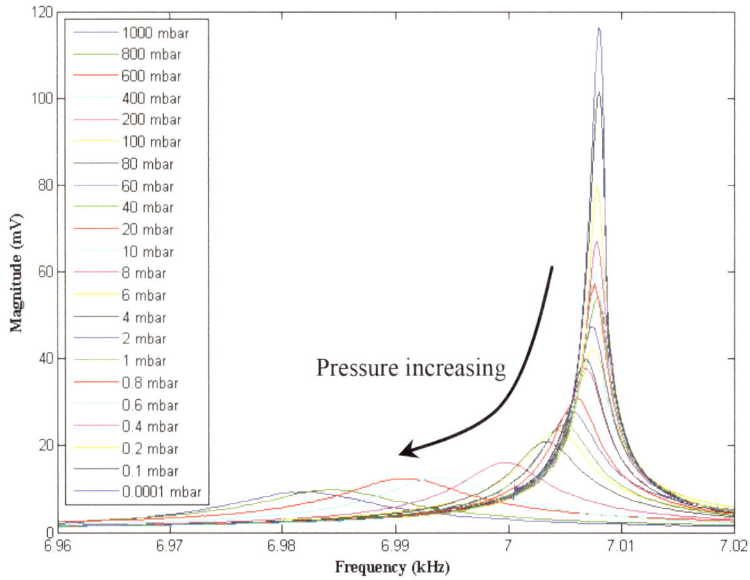

Figure 5. Resonant measurements of a microresonator under the pressures from high vacuum to 1000 mbar in an N_2 environment for the first flexural mode.

$$\partial \mathbf{u}/\partial t + (\mathbf{u} \cdot \nabla)\mathbf{u} = -\frac{1}{\rho}\nabla p + \frac{4}{3}\frac{\eta}{\rho}\nabla(\nabla \cdot \mathbf{u}) - \frac{\eta}{\rho}\nabla \times (\nabla \times \mathbf{u}), \tag{1}$$

$$\rho(\nabla \cdot \mathbf{u}) + \partial\rho/\partial t = 0, \tag{2}$$

$$p = \left(C_p - C_v\right)\rho T, \tag{3}$$

$$\rho C_p \partial T/\partial t = \kappa \Delta T + \partial p/\partial t, \tag{4}$$

where $\mathbf{u}$, C_v, and t denote, respectively, the velocity vector, heat capacity at constant volume, and time. The operators ∇ and Δ are the gradient and Laplace operator, respectively.

A solution of the full model called Boundary Element Method (BEM) [21, 22] is discussed here. The velocity is written as the sum of a viscous velocity $\mathbf{u}_v$, due to viscous effects, and a laminar velocity $\mathbf{u}_l$:

$$\mathbf{u} = \mathbf{u}_v + \mathbf{u}_l, \tag{5}$$

which satisfy the conditions that the divergence of the viscous velocity is zero: $\nabla \cdot \mathbf{u}_v = 0$, and the rotation of the laminar velocity is zero: $\nabla \times \mathbf{u}_l = 0$.

The pressure is also split up into two components:

$$p = p_a + p_h, \tag{6}$$

where p_a is the acoustic pressure and p_h is the thermal pressure. Splitting the acoustic variables facilitates rewriting the governing equations into scalar wave equations for the acoustic and thermal pressures and a vector wave equation for the viscous velocity:

$$(\Delta + k_a^2)p_a = 0, \tag{7}$$

$$(\Delta + k_h^2)p_h = 0, \tag{8}$$

$$(\Delta + k_v^2)\mathbf{u}_v = 0. \tag{9}$$

The temperature fluctuation ΔT is the sum of acoustic and thermal temperature variations related to the acoustic and thermal pressures by:

$$\Delta T = \alpha_a p_a + \alpha_h p_h, \tag{10}$$

and the laminar velocity $\mathbf{u}_l$ is written as:

$$\mathbf{u}_l = \phi_a \nabla p_a + \phi_h \nabla p_h. \tag{11}$$

The exact expressions for k_a, k_h, k_v, α_a, α_h, ϕ_a, and ϕ_h can be found in the literatures [21, 22].

3.1.2. Physical interpretation

The solution of the full N-S model can be interpreted as follows: the resonator vibration results in acoustic wave propagation in the ambient fluid. The acoustic domain can be divided into a boundary layer and the bulk region, as illustrated in **Figure 6**. In microscaled geometries, the boundary layer occupies a substantial part of the acoustic domain, the acoustic model needs to account for the viscothermal effects to accurately describe the wave propagation.

Mechel [23] describes the viscothermal acoustic equations as the interaction of viscous, thermal, and acoustic waves. The viscous k_v, thermal k_h, and acoustic k_a wave numbers are derived as: $k_v^2 = -i\omega\rho/\eta$, $k_h^2 = -i\omega\rho C_p/\kappa$, and $k_a = \omega/c_0$. Viscous and thermal waves are heavily damped, as their wavelengths λ_v and λ_h have comparable length scales with the boundary layer

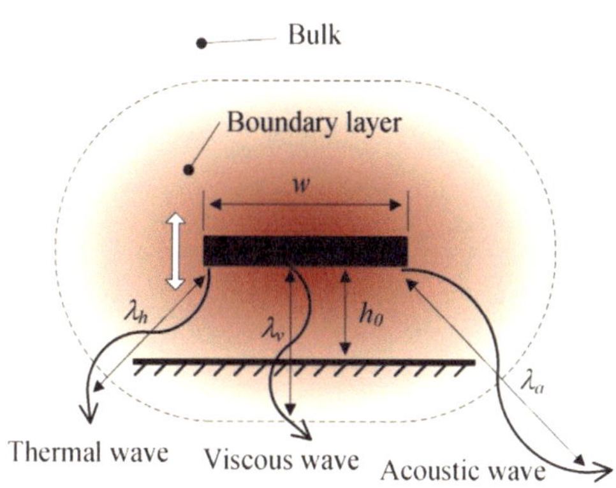

Figure 6. Two regions in the acoustic domain: the bulk and the boundary layer, and three waves due to vibration.

thickness; while the acoustic wave is slightly damped and propagates mainly in the bulk regime. **Figure 7** compares these characteristic length scales for vibration in air at frequencies from 1 kHz to 1 MHz.

For microresonators vibrating at low frequencies, the characteristic length of the resonator (e.g., width of the resonant beam) is much smaller than the acoustic wavelength propagating in the fluid. The acoustic wave develops so weakly that the bulk region can be ignored and the fluid is assumed as incompressible, and the vibration energy is mainly dissipated due to the viscothermal effects. Consequently, the fluid motion can be modeled accurately by incompressible N-S equations. However, as frequency increases, the acoustic wavelength reduces to smaller than the characteristic length of the resonator. The bulk regime can become significant and the acoustic wave starts to radiate vibration energy. Therefore, compressibility can become important for operation at higher frequencies.

3.1.3. Comparison between experiments and simulations

Resonators with different geometry and dimensions have been fabricated to verity the predictions of the full N-S model. One of the AlN-based resonators is shown in **Figure 8**, the insert shows the resonator chip mounted on a printed circuit board. The resonator has been tested from high vacuum to normal atmosphere.

The full model equations are solved and simulated with a self-developed FEM solver using MATLAB. **Figure 9** shows the simulated pressure, density, temperature, and velocity amplitude distributions of the air gap in the *xz* plane at 0.01 and 1000 mbar. The amplitude profiles in **Figure 9** indicate that, compare to vibration in the higher pressure, the pressure and density perturbation are more constant, and the temperature is more homogenous across the air gap in the lower pressure.

The measured and simulated air damping coefficients under different pressures are shown in **Figure 10**. This figure indicates that simplified models, such as viscous model or squeeze model, are not suitable in this case, since they are accurate only for resonators with slim beams

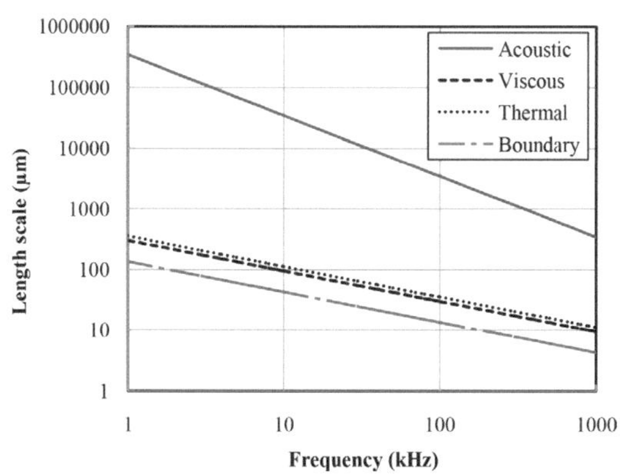

Figure 7. Characteristic length scales for vibrations in air: acoustic (solid), viscous (dashed), thermal (dotted) wavelengths, and boundary layer thickness (dash-dotted).

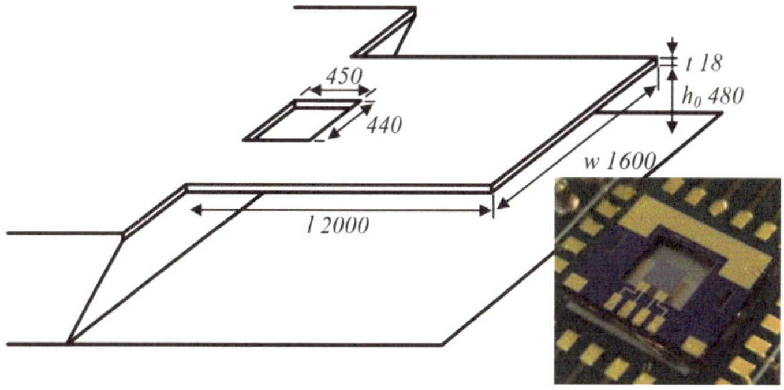

Figure 8. The geometry of the resonator (in μm). The insert shows the photo of the resonator [24].

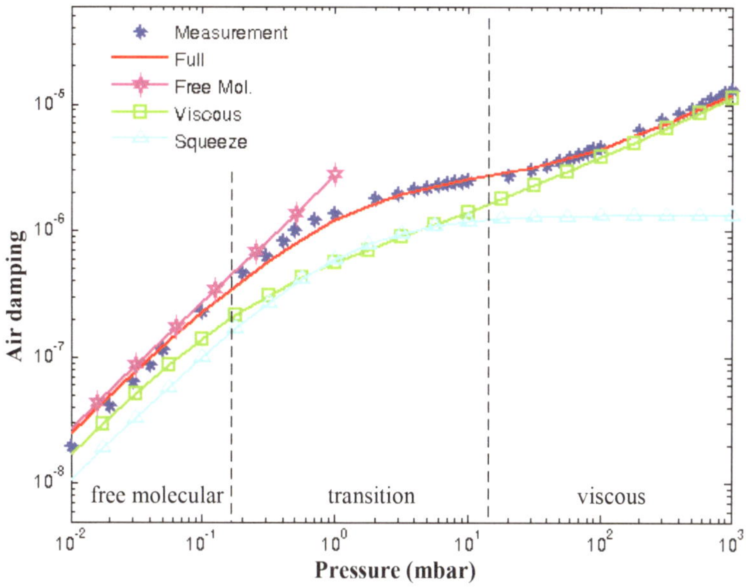

Figure 9. Amplitude profiles of (a) pressure, (b) density, (c) temperature, and (d) velocity in the air gap at 0.01 mbar (column I) and at 1000 mbar (column II) [24].

or narrow air gaps. In the free molecular region, the free molecular model is reasonable. The full N-S model yields good agreement in the whole pressure range.

3.2. Hydrodynamics of microresonators vibrating in unbounded fluids

3.2.1. Measurement results at atmospheric pressure

The AlN-based microresonators were tested in the chamber filled with different gases under atmospheric pressure, as shown in **Figure 11**. The resonance curve variations indicate that the resonance frequency f_r decreases as the density of the gas increases, shifting from 7 kHz for He (ρ = 0.18 kg/m^3) to 6.91 kHz for Xe (ρ = 5.90 kg/m^3).

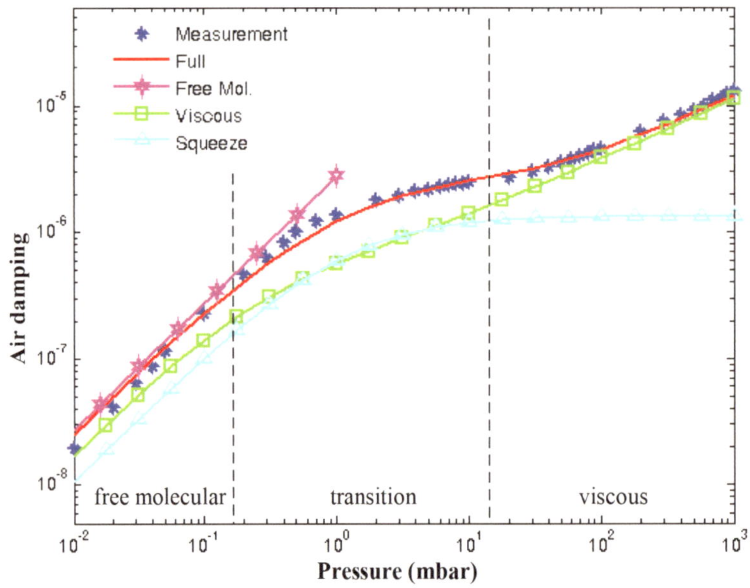

Figure 10. Air damping coefficient of the resonator as a function of ambient pressure [24].

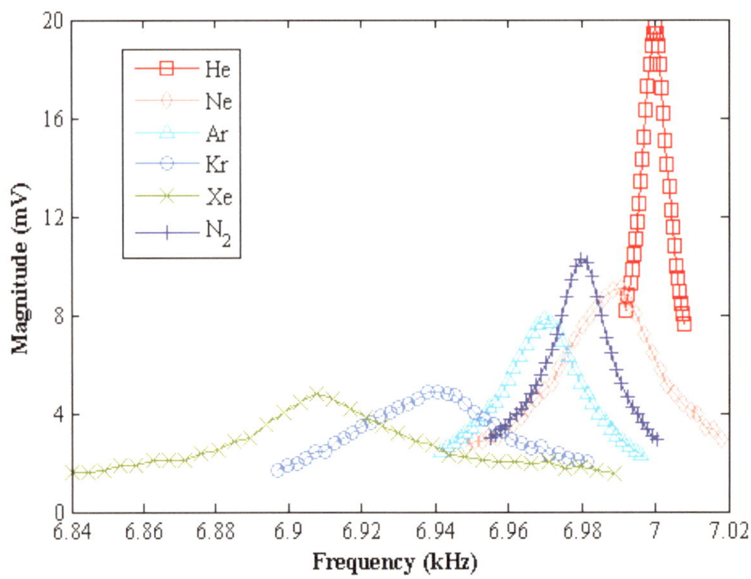

Figure 11. Resonant measurements of the cantilever with different gases at atmospheric pressure in its first flexural mode [25].

To excite the cantilever in higher resonant modes, the frequency of the driving voltage was scanned from 5 to 250 kHz. **Figure 12** shows the resonance frequency response of the cantilever in N_2 at atmospheric pressure. Finite element analysis software COMSOL was used to assign the resonant modes to the observed peaks. The resonant mode shapes inserted in the figure are obtained by this software and depict the displacement of the cantilever for each mode.

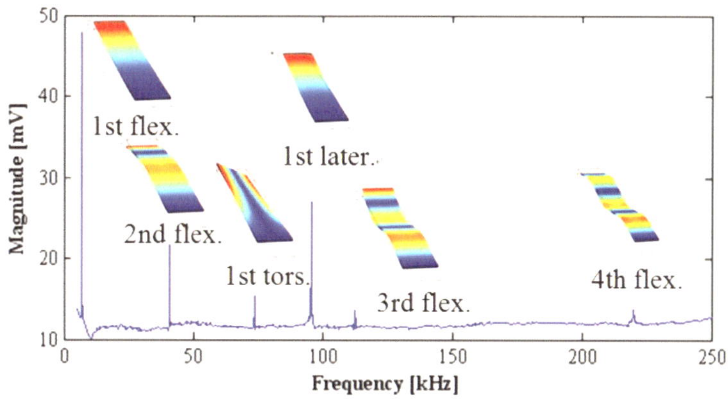

Figure 12. Amplitude spectrum and mode shapes of the cantilever at normal atmospheric N_2 [25].

To evaluate the fluidic hydrodynamic loading effects, the cantilever vibrating in different resonant modes has been characterized in different gases at atmospheric pressure. Only the first two modes for Kr and Xe could be detected, since the signal of higher modes was too weak. The remaining gases are successfully characterized up to the fourth flexural mode. Pure fluidic hydrodynamic loading is evaluated by subtracting the intrinsic damping, which can be measured in a high vacuum.

The quality factor induced by gas damping Q_{gas} in atmospheric pressure is characterized and illustrated in **Figure 13**. Resonator immersed in He shows the highest Q factors for all modes. The two lightest gases, He and Ne, exhibit a continuous increase in Q_{gas} with the mode number, whereas for Ar and N_2, Q_{gas} decreases beyond the third mode. Taking N_2 as an example, the measured values of Q_{gas} increase from 543 to 1890 for the first to third

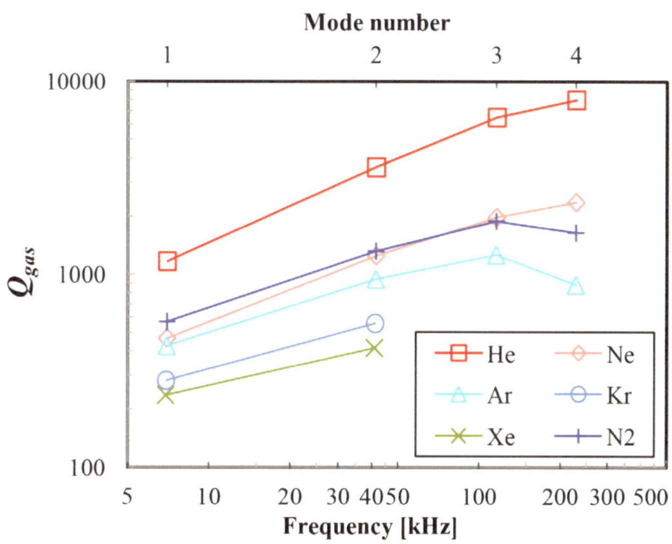

Figure 13. Q factors due to gas damping of the different flexural modes at atmospheric pressure [25].

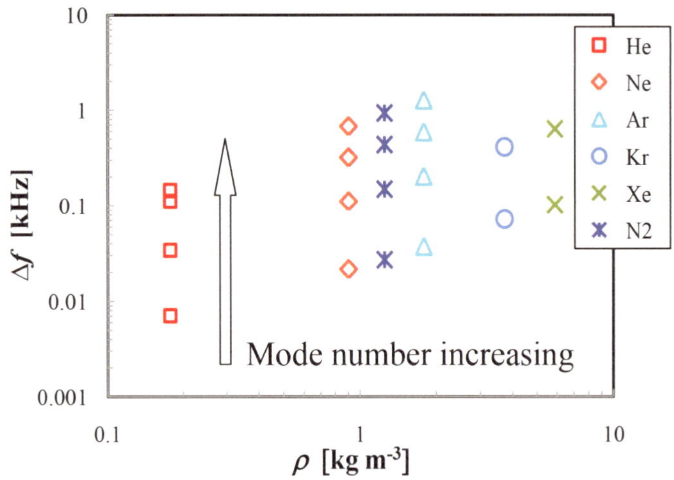

Figure 14. Resonance frequency shifts Δf of the different flexural modes at atmospheric pressure [25].

modes. However, this trend is broken when Q_{gas} reduces to 1645 for the fourth flexural mode, indicating that Q_{gas} will not increase in an unlimited way with increasing mode/frequency.

The corresponding resonance frequency shift Δf, defined as $f_{r,vac} - f_r$, where $f_{r,vac}$ corresponds to the resonance frequency f_r in vacuum, has been characterized. **Figure 14** shows Δf as a function of density ρ for different gases. The results in this figure indicate a dependence on both f_r and ρ.

3.2.2. Measured results at reduced pressures

The resonators were further tested at reduced pressures to investigate the influence of ambient pressure. **Figure 15** shows the measured Q factor curve of one resonator vibrating at the first flexural mode as a function of N_2 pressure. When the pressure is sufficiently low ($<10^{-3}$ mbar in this figure), the damping is independent of the fluid and relates only with intrinsic losses in the resonator structure. Consequently, this pressure regime is called the intrinsic regime. Gas damping starts to become visible at higher pressure levels. Three regions can be identified by using the Knudsen number Kn (defined as the ratio between the mean free path of the fluid and the width of the resonator):

- The molecular regime ($Kn > 10$): damping is caused by independent collisions of non-interacting fluid molecules with the vibrating surface of the resonator and/or surrounding walls.

- The transition regime ($0.01 < Kn < 10$): the fluid is neither noninteracting nor continuous.

- The continuous regime ($Kn < 0.01$): the fluid acts as continuum and most previous papers (e.g., [26, 27]) found that the viscous drag is typically the dominant energy loss mechanism based on the incompressible gas assumption.

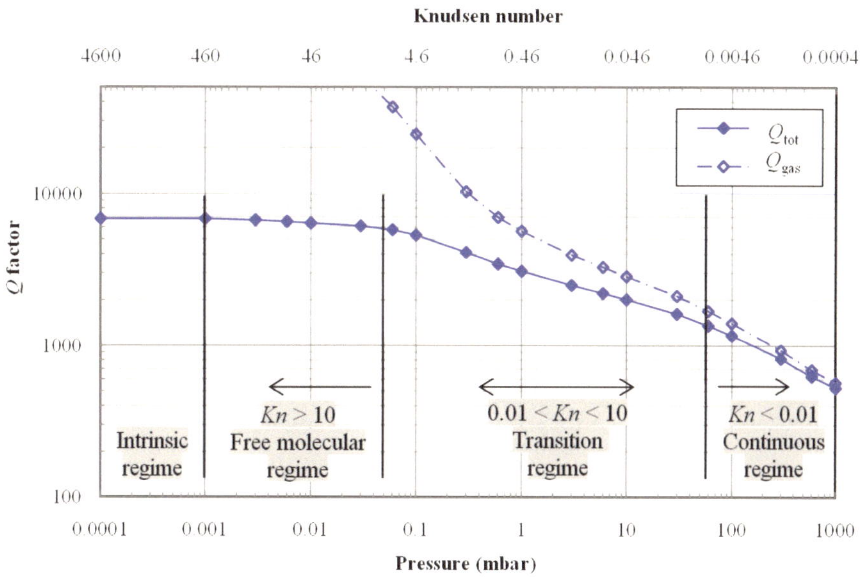

Figure 15. Variation of the Q factor with N_2 pressure. The solid diamond points are the measurements, while the hollow points are the derived Q factors due to fluidic damping.

3.2.3. Results analysis

It is found that the fluidic loading has a strong effect on the resonant behavior of the resonator, especially at higher pressures (e.g., near atmospheric pressure). The following analysis focuses on fluidic damping at atmospheric pressure, which is commonly present in many practical applications.

Viscous drag is typically the dominant loss mechanism, for most microdevices vibrating at relatively low frequency. For a microcantilever with dimensions of length l, width w, and thickness h, undergoing flexural vibrations in a continuous incompressible fluid, the corresponding Q factor due to viscous damping is given by [25, 28, 29]:

$$Q_{vis} \approx \frac{\rho_b h}{\sqrt{\pi}} \sqrt{\frac{f_r}{\rho \eta}}, \tag{12}$$

where ρ_b is the cantilever beam density. It can be seen that the Q factor scales inversely proportional with $\sqrt{\rho\eta}$ and proportional with $\sqrt{f_r}$, which means that a higher resonance frequency (or higher resonant mode) results in a higher Q factor. The measured and calculated Q factors Q_{gas} of the different flexural modes are plotted in **Figure 16** as a function of the combination of the resonance frequency f_r, the density ρ, and the viscosity η of the surrounding gases. However, as noted before, a decrease of the quality factor is observed for the third and fourth modes of Ar and N_2.

One possible reason is the additional acoustic damping due to the compression of the fluid. Incompressible flow is expected for low frequency vibrations, since the wavelength of sound in the fluid is much longer than the dominant length scale of the vibrating beam. As the resonant mode increases, the acoustic wavelength reduces and the incompressible gas assumption is no

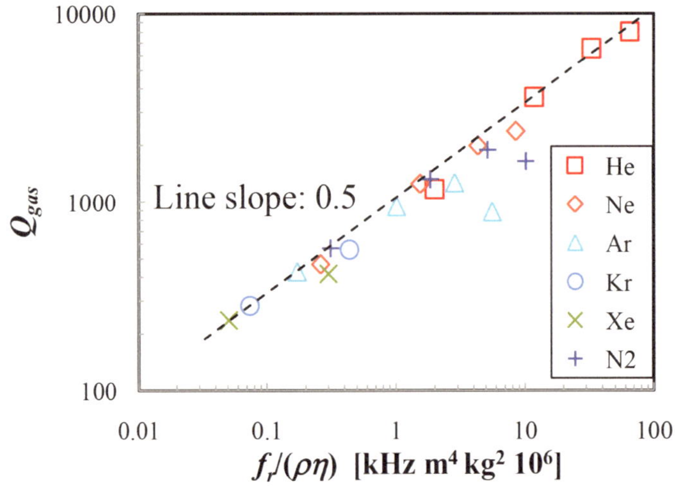

Figure 16. Evolution of the Q factors as a function of $f_r/(\rho\eta)$ for different gases and resonant modes at atmospheric pressure. The dashed line represents the viscous model (Eq. (1)) prediction [25].

longer valid [25, 29, 30]. Thus, at high frequencies, acoustic energy loss becomes important in addition to viscous loss.

3.3. Hydrodynamics of microresonators vibrating close to a surface

3.3.1. Measured results

We have measured bridge resonator different gap depths to understand how fluidic hydrodynamic load is modified when the resonator is vibrating near a surface. The case of vibration at free space was also measured for comparison. We begin by presenting results of the resonance responses of the resonator immersed in atmospheric N_2 at room temperature, as shown in **Figure 17**.

It is obvious that the air gap has a strong effect on the resonance behavior. A general trend is the resonance peak getting broaden and shifting to lower peak frequency as the gap depth decreases. There results fit with the expectation that the fluidic damping increases significantly when approaching toward a surface. We have also observed that the resonance frequency shift is relatively insensitive to air gap when the separation h_0 is bigger than 150 μm, which corresponds to a gap to the resonator width ratio of more than 0.5. However, for h_0 = 150 μm, the Q factor is approximately 20% less than vibration without gap, and for h_0 = 250 μm, the Q factor differs by only less than 2%.

To quantify the pressure effect on the dissipation in the fluid as the resonator is vibrating near a surface, it was further measured under reduced pressures. **Figure 18** shows the quality factors for different gap heights and pressures in N_2 atmosphere.

3.3.2. Review of analytical models

In many practical applications, restrictive squeeze-film assumption (very narrow air gap) does not always hold; thus, traditional squeeze-film models are not suitable in those cases. Two

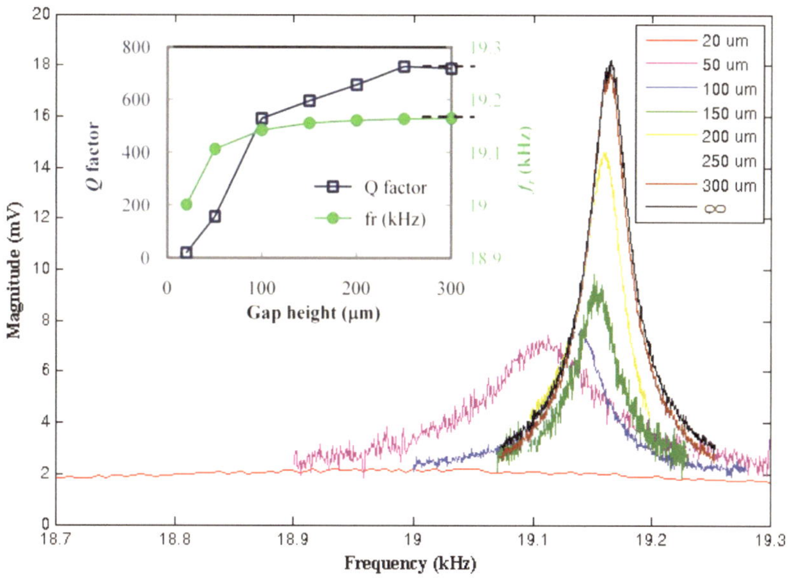

Figure 17. Resonant curves of a bridge resonator in fundamental mode at atmospheric N_2. The curves for gap depths of 250 and 300 μm and ∞ are overlapped. The dashed lines in the inserted figure represent the limit for the resonator at free space [31].

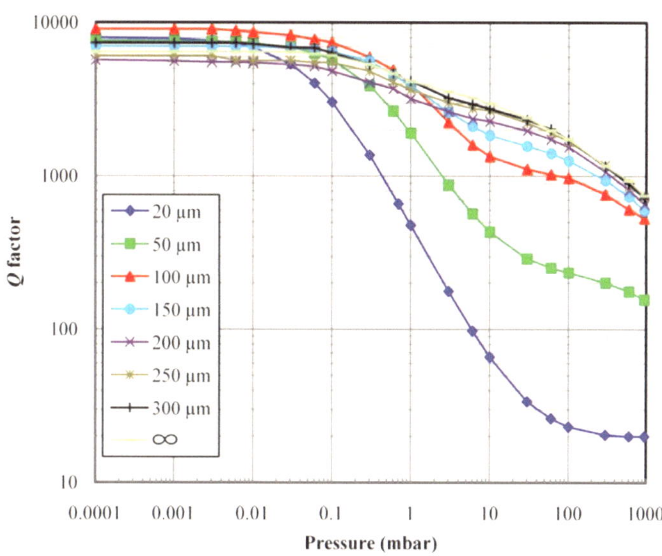

Figure 18. Variation of Q factors with ambient pressure for N_2 with different gap heights [31].

recently improved models, namely (a) extended squeeze-film model [32, 33] and (b) unsteady N-S model [34–36], are considered for describing the behavior of the resonators with moderate gap depths.

3.3.2.1. Extended squeeze-film model

Osborne Reynolds first formulated the theory for a squeezed film between two surfaces in relative motion to each other more than a century ago [37]. Generally, the compressibility should be considered and its importance increases as vibration frequency. An important measure for the squeezed-film effect is the squeeze number:

$$\sigma = \frac{12\eta w^2 \omega}{p{h_0}^2}, \tag{13}$$

where η is the fluid viscosity, ω is the angular frequency of the resonator, p is the ambient pressure, w is the width of the resonator, and h_0 is the gap depth. The number σ specifies the ratio between the spring force due to the gas compressibility and the force due to the viscous flow. When σ is much smaller than 1, the gas in the film has enough time to "leak" out; thus, the gas is referred to as incompressible. For the resonator geometry in this study, the resonator width w and the gap depth h_0 are of the same order, e.g., $h_0/w > 0.05$. For air at atmospheric pressures, the vibrating frequency should be higher than 150 kHz to ensure σ is bigger than 1.

When σ is very small, the compressible effects can be ignored and the damping coefficient for a slender resonator can be written as:

$$c = \frac{\eta l w^3}{{h_0}^3}, \tag{14}$$

and Q factor is then given by:

$$Q = \frac{\rho_b t \omega}{\eta w^2} {h_0}^3, \tag{15}$$

herein ρ_b, l and, t are the density, length, and thickness of the resonator, respectively.

An effective plate width $w_{eff} = w + \Delta w$ is introduced, to include border effects into the much simplified squeeze-film model. The effective width is such that the damping force for the enlarged plate with trivial boundary conditions has the same values as that of a real device size with the border effects. This concept is illustrated in **Figure 19**.

One method to predict the elongation caused by the border effect was proposed by Veijola et al. [32, 33], based on a series of 2D and 3D FEM simulations. A simple conclusion has been obtained for a slender beam:

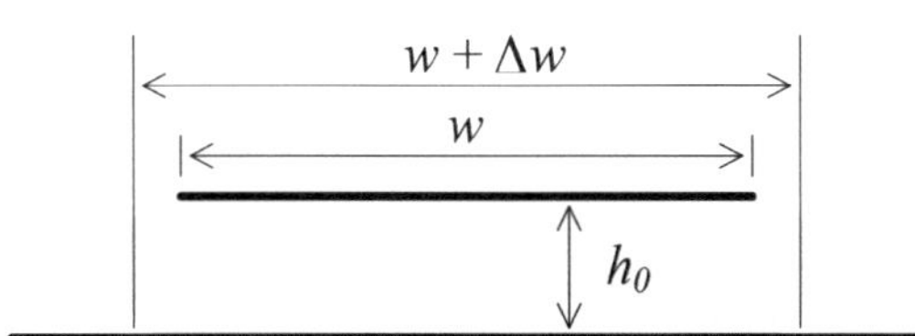

Figure 19. Schematic diagram of extended squeeze-film model.

$$\Delta \mathrm{w} = 1.3h_0. \tag{16}$$

The fluidic damping coefficient due to air gap is:

$$c = \frac{\eta l (w + 1.3h_0)^3}{h_0{}^3}. \tag{17}$$

The quality factor consequently is:

$$Q = \frac{\rho_b w t \omega}{\eta (w + 1.3h_0)^3} h_0{}^3. \tag{18}$$

To account the non-continuum fluid behavior at very low pressures or for a narrow gap, an effective viscosity η_{eff} is used instead of the gas dynamic viscosity η. Burgdoufer [38] obtained a simple form for the effective viscosity coefficient:

$$\eta_{eff} = \frac{\eta}{1 + 6K_n}. \tag{19}$$

3.3.2.2. Unsteady N-S model

Another method to deal with moderate gaps can be based on the unsteady incompressible N-S equation and the continuity equation [20]. For a beam undergoing normal oscillations in a viscous fluid, the general form of the hydrodynamic force in the vibration direction is given as [39]:

$$\mathbf{F_d} = \frac{\pi}{4} \rho \omega^2 w^2 l \Gamma(Re) \hat{W}(x|\omega), \tag{20}$$

where ρ is the fluid density, $\hat{W}$ is the Fourier transformation of the resonator beam deformation. Γ, the so-called "hydrodynamic function", is a complex term: the real part Γ_{real} represents the inertial forces of the fluid or added mass components, whereas the imaginary part Γ_{imag} is proportional to the viscous forces of the fluid or damping components.

The general form of the semi-analytical formula for the hydrodynamic function is $\Gamma(Re, H) = 10^{\Gamma_L}$ where

$$\begin{aligned} \Gamma_L(Re, H_L) = {} & a_1 + a_2 Re_L + a_3 Re_L{}^2 + a_4 Re_L{}^3 + a_5 Re_L{}^4 \\ & + a_6 Re_L H_L + a_7 H_L + a_8 H_L{}^2 + a_9 H_L{}^3 + a_{10} H_L{}^4 \\ & + a_{11} Re_L H^L L^2 + a_{12} Re_L{}^2 H_L + a_{13} Re_L H_L{}^3 + a_{14} Re_L{}^3 H_L \\ & + a_{15} (Re_L H_L)^2 + a_{16} (Re_L H_L)^3, \end{aligned} \tag{21}$$

$H_L = \log_{10}(H)$, and $Re_L = \log_{10}(Re)$. The coefficients a_k, $k = 1, 2, \ldots, 16$, have complex values. They are tabulated in Tung et al.'s paper [36]. The fit is valid in the range of $10^{-2} < Re < 10^4$ and $10^{-1} < H < 10$. The Q factor due to the fluid damping is defined by

$$Q = \frac{4\rho_b t}{\pi\rho w \Gamma_{imag}}. \tag{22}$$

The calculations are technically valid only when the continuum hypothesis holds. This is because the theory is based upon a subset of the N-S equations, which are based on a continuum assumption. The useful range of this semi-analytical expression can be extended beyond the continuum regime and into the slip and transition regimes by using an effective viscosity concept.

3.3.3. Results analysis

The predictions from squeeze film (Eq. (13)), extended squeeze film (Eq. (18)), unsteady N-S (Eq. (22)), and viscous (Eq. (12)) models are compared with the experimental data. **Figure 20** shows the results for the first flexural mode of the bridge resonator oscillating in atmospheric N_2 with different gap depths. The intrinsic damping of the resonator can be obtained by operating in a high vacuum, and this damping was subsequently removed mathematically, leaving only the fluidic damping to be analyzed.

The traditional squeeze-film damping model overestimates the Q factor and the divergence increases rapidly as the gap depth increases. While the extended squeeze-film model predicts well the squeeze-film damping up to gap depth of 50 μm. This model accurately predicts the fluidic damping well even into the free molecular regime, indicating the "effective viscosity" concept is reasonable to account the rarefaction effect of the gas in a narrow gap. The unsteady N-S model predictions are in better agreement with measurements as compared to other models. The viscous model fits well with the measurements in the viscous regime, but when the pressure decreases into the transition regime, this model loses its validity.

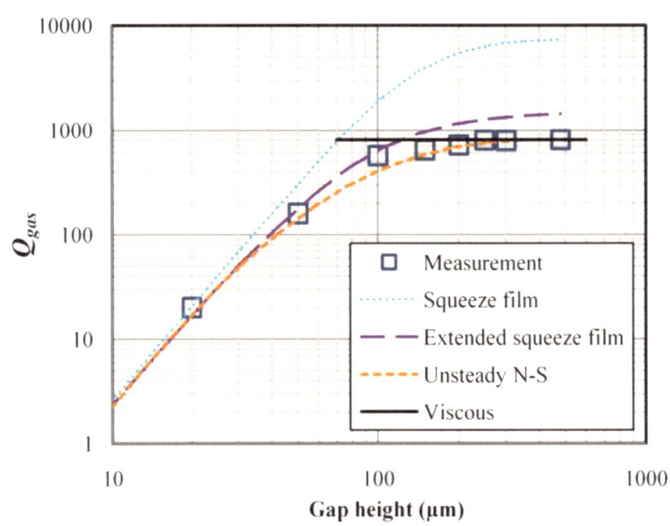

Figure 20. Q factors due to fluidic damping of a resonator. The squares represent the measurements, and the dashed lines represent the analytical/semianalytical calculations, while the solid line shows the viscous model prediction [31].

http://dx.doi.org/10.5772/intechopen.77731

4. Conclusion and discussion

In this work, AlN-based piezoelectric microresonators have been fabricated and measured to study the fluidic hydrodynamic loading mechanisms. Experimental, computational, and analytical tools have been developed, to analyze the dynamics of the resonators in different gases from atmospheric pressure to high vacuum.

For resonators vibrating in a continuous fluid, the extensive type of theoretical model is based on the full set of N-S equations. The equations were semi-analytically solved using the BEM method. The solution can be interpreted as coupling of viscous, thermal, and acoustic waves. However, this method does have some disadvantages, such as being computationally costly, relying on sophisticated numerical techniques, and being non-intuitive with respect to understand the results physically.

The resonators vibrating in free space fluids have been investigated, by measuring microcantilevers in different gases and explaining the results using simplified models based on the incompressible N-S equation. The analytical models agree reasonably with measurements. Besides, it is found that the compressibility of gases leads to additional acoustic damping when vibrating in higher mode/frequency.

For vibration close to a surface, experiments were performed using a bridge resonator with a big range of gap depths. The traditional squeeze-film model could predict the squeeze-film damping when the ratio of the resonator width to gap height w/h_0 was bigger than 10, and the gas rarefaction effect in the gap can be accounted by using an "effective viscosity" concept. The extended squeeze-film model was valid for $w/h_0 \geq 4$. The unsteady N-S model could be used when the gap height was even bigger.

Author details

Huacheng Qiu[1]* and Helmut Seidel[2]

*Address all correspondence to: huacheng.qiu@outlook.com

1 China Aerodynamics Research and Development Center, Hypervelocity Aerodynamics Institute, Sichuan Province, China

2 Saarland University, Chair of Micromechanics, Microfluidics/Microactuators, Saarbrücken, Germany

References

[1] Ruby R. Review and comparison of bulk acoustic wave FBAR, SMR technology. In: Proc. IEEE Ultrasonic Symposium. 2007. pp. 1029-1040

[2] Martin PM, Good MS, Johnston JW, Posakony GJ, Bond LJ, Crawford SL. Piezoelectric films for 100-Mhz ultrasonic transducers. Thin Solid Films. 2000;**379**:253-258

[3] Madni AM, Costlow LE, Knowles SJ. Common design techniques for BEI GyroChip quartz rate sensors for both automotive and aerospace/defense markets. IEEE Sensors Journal. 2003;**3**:569-578

[4] Beeby SP, Ross JN, White NM. Design and fabrication of a micromachined silicon accelerometer with thick-film printed PZT sensor. Journal of Micromechanics and Microengineering. 2000;**10**:322-328

[5] Stephanou PJ, Piazza G, White CD, Wijesundara MBJ, Pisano AP. Piezoelectric aluminum nitride MEMS annular dual contour mode filters. Sensors and Actuators A. 2007;**134**:152-160

[6] Horowitz SB, Sheplak M, Cattafesta LN, Nishida T. A MEMS acoustic energy harvester. Journal of Micromechanics and Microengineering. 2006;**16**:174-181

[7] Sökmen Ü, Stranz A, Waag A, Ababneh A, Seidel H, Schmid U, Peiner E. Evaluation of resonating Si cantilevers sputter-deposited with AlN piezoelectric thin films for mass sensing applications. Journal of Micromechanics and Microengineering. 2010;**20**:064007

[8] Ayela C, Nicu L. Micromachined piezoelectric membranes with high nominal quality factors in Newtonian fluid media: A lamb's model validation at the microscale. Sensors and Actuators B: Chemical. 2007;**123**:860-868

[9] James D, Scott SM, Ali Z, O'Hare WT. Chemical sensors for electronic nose systems. Microchimica Acta. 2005;**149**:1-17

[10] Waggoner PS, Craighead HG. Micro- and nanomechanical sensors for environmental, chemical, and biological detection. Lab on a Chip. 2007;**7**:1238-1255

[11] Fadel L, Lochon F, Dufour I, Français O. Chemical sensing: Millimeter size resonant microcantilever performance. Journal of Micromechanics and Microengineering. 2004;**14**:S23

[12] Kim SJ, Ono T, Esashi M. Study on the noise of silicon capacitive resonant mass sensors in ambient atmosphere. Journal of Applied Physics. 2007;**102**:104304

[13] Xia X, Zhang Z, Li X. A Latin-cross-shaped integrated resonant cantilever with second torsion-mode resonance for ultra-resoluble bio-mass sensing. Journal of Micromechanics and Microengineering. 2008;**18**:035028

[14] Zhou J, Li P, Zhang S, Huang Y, Yang P, Bao M, Ruan G. Self-excited piezoelectric microcantilever for gas detection. Microelectronic Engineering. 2003;**69**:37-46

[15] Ababneh A, Schmid U, Hernando J, Sánchez-Rojas JL, Seidel H. The influence of sputter deposition parameters on piezoelectric and mechanical properties of AlN thin films. Materials Science and Engineering B. 2010;**172**:253-258

[16] Qiu HC, Ababneh A, Feili D, Wu XZ, Seidel H. Analysis of intrinsic damping in vibrating piezoelectric microcantilevers. Microsystem Technologies. 2016;**22**(8):2017-2025

[17] Hernando J, Sánchez-Rojas JL, González-Castilla S, Iborra E, Ababneh A, Schmid U. Simulation and laser vibrometry characterization of piezoelectric AlN thin fims. Journal of Applied Physics. 2008;**104**:053502

[18] Qiu HC, Schwarz P, Feili D, Merzsh S, Peiner E, Wu XZ, Seidel H. Electrical performance analysis and characterization of two port piezoelectric resonators. Microsystem Technologies. 2013;**19**:1131-1136

[19] Qiu HC, Schwarz P, Völlm H, Feili D, Wu XZ, Seidel H. Electrical crosstalk in two-port piezoelectric resonators and compensation solutions. Journal of Micromechanics and Microengineering. 2013;**23**:045007

[20] Landau LD, Lifshitz EM. Fluid Mechanics. 2nd ed. Oxford: Pergamon Press; 2004

[21] Beltman WM. Viscothermal wave propagation including acousto-elastic interaction, part I. Theory. Journal of Sound and Vibration. 1999;**227**:555-586

[22] Beltman WM. Viscothermal wave propagation including acousto-elastic interaction, part II. Applications. Journal of Sound and Vibration. 1999;**227**:587-609

[23] Mechel FP. Revision of the Kirchhoff-Rayleigh theory of sound propagation in viscothermal air. Acta Acustica United with Acustica. 2007;**93**:507-534

[24] Qiu HC, Schwarz P, Feili D, Wu XZ, Seidel H. Viscothermal acoustic waves in micro scale resonators. In: Proceedings of Transducers 2013 & Eurosensors XXVII. 2013. 1719-1722

[25] Qiu HC, Xiao DB, Feili D, Wu XZ, Seidel H. Hydrodynamic analysis of piezoelectric microcantilevers vibrating in viscous compressible gases. Sensors and Actuators A. 2016; **238**:299-306

[26] Johnson BN, Mutharasan R. Dependence of the quality factor of micromachined silicon beam resonators on pressure and geometry. Journal of Vacuum Science and Technology B. 1992;**10**:19-26

[27] Bhiladvala RB, Wang ZJ. Effect of fluids on the *Q* factor and resonance frequency of oscillating micrometer and nanometer scale beams. Physical Review E. 2004;**69**:036307

[28] Blom FR, Bouwstra S, Elwenspoek M, Fluitman JHJ. Dependence of the *Q* factor of micromachined silicon beam resonators on pressure and geometry. Journal of Vacuum Science and Technology. 1992;**10**:19-26

[29] Qiu HC, Feili D, Wu XZ, Seidel H. Resonant-mode effect on fluidic damping of piezoelectric microcantilevers vibrating in an infinite viscous fluid. Sensors and Actuators A. 2014; **232**:1-7

[30] Van Eysden CA, Sader JE. Compressible viscous flows generated by oscillating flexible cylinders. Physics of Fluids. 2009;**21**:013104

[31] Qiu HC, Schwarz P, Feili D, Wu XZ, Seidel H. Hydrodynamics of micro beam resonator vibrating close to a surface with a moderate distance. Journal of Micromechanics and Microengineering. 2014;**25**:055016

[32] Veijola T, Pursula A, Raback P. Extending the validity of existing squeeze-film damper models with elongations of surface dimensions. In: Proc. Nanotech 2004. 2004. pp. 235-238

[33] Veijola T, Pursula A, Raback P. Extending the validity of squeeze-film damper models with elongations of surface dimensions. Journal of Micromechanics and Microengineering. 2005; **15**:1624-1636

[34] Green CP, Sader JE. Frequency response of cantilever beams immersed in viscous fluids near a solid surface with applications to the atomic force microscope. Journal of Applied Physics. 2005;**98**:114916

[35] Green CP, Sader JE. Small amplitude oscillations of a thin beam immersed in a viscous fluid near a solid surface. Physics of Fluids. 2005;**17**:073102

[36] Tung RC, Jana A, Raman A. Hydrodynamic loading of microcantilevers oscillating near rigid walls. Journal of Applied Physics. 2008;**104**:114905

[37] Langlois WE. Isothermal squeeze films. Quarterly of Applied Mathematics. 1962;**XX**(2): 131-150

[38] Burgdorfer A. The influence of the molecular mean free path on the performance of hydrodynamic gas lubricated bearings. Journal of Basic Engineering. 1959;**81**:94-99

[39] Sader JE. Frequency response of cantilever beams immersed in viscous fluids with applications to the atomic force microscope. Journal of Applied Physics. 1998;**84**:64-76

Chapter 4

Piezoelectric Melt-Spun Textile Fibers: Technological Overview

Dimitroula Matsouka and Savvas Vassiliadis

Additional information is available at the end of the chapter

http://dx.doi.org/10.5772/intechopen.78389

Abstract

Piezoelectricity was first described by the Curie brothers in the late 1800s. The first materials investigated were natural materials such as bone and wood and single crystals such as quartz. Then in 1946 it was discovered that $BaTiO_3$ ceramic can be made piezoelectric through a poling process. This was followed by the discovery of lead zirconate titanate solid solutions (PZT) in 1954 of very strong lead effects which is still widely used in piezoelectric applications. In 1969, Kawai discovered large piezoelectricity in elongated and poled films of polyvinylidene fluoride (PVDF) opening the way for research into piezoelectric polymers. Piezoelectric polymers exhibit low density and excellent sensitivity and are mechanically tough and respond better to fatigue situations. Since 2010, research has focused on the production of melt-spun piezoelectric textile fibers, with the aim of integrating sensing/energy-harvesting capabilities into smart textile structures. In this chapter, a technological overview of the state-of-the-art research into piezoelectric, melt-spun, textile fibers will be presented. The methods used for the characterization of the fibers will also be discussed with special concentration on the electric response of the fibers after mechanical stimulation.

Keywords: piezoelectricity, textiles, melt spinning

1. Introduction

Piezoelectricity, discovered in 1880 by Pierre and Jacques Curie in quartz [1], is observed in all materials with a crystalline anisotropy. Piezoelectricity has two distinct effects. The direct effect is the polarization of the material under mechanical stress and the inverse effect corresponds to a mechanical displacement when electric polarization is applied to the material [2].

Piezoelectric materials belong to the general group of dielectric materials, electrical insulators that can be polarized by an applied electric field (**Figure 1**). Piezoelectric materials are non-centrosymmetric dielectrics; this means that when subjected to an external electric field, there will be asymmetric movement of the neighboring ions, resulting in significant deformation of the structure; this deformation is directly proportional to the applied electric field [3].

Pyroelectricity, the ability of certain materials to generate an electrical potential when they are heated or cooled, occurs in all materials that belong to a polar crystal symmetry class. It should be noted that, not all non-centrosymmetric classes are polar, not all piezoelectric crystals are pyroelectric. However, all pyroelectric crystals are piezoelectric. Ferroelectrics form a subset of the set of pyroelectrics because they are polar materials in which the direction of the polar axis can be changed by the application of an electric field [4]. Investigation into the piezoelectric properties of materials commenced from materials readily available in nature such as carnauba wax [5], wood [6] and bone [7]. In 1946, it was shown that $BaTiO_3$ ceramic can be made piezoelectric by an electrical poling process. The first commercial piezoelectric devices based on $BaTiO_3$ ceramics were phonograph pickups and appeared in the market in about 1947 [8]. An advance of great practical importance was the discovery in 1954 of very strong piezoelectric effects in lead zirconate titanate solid solutions (PZT) [9]. PZT piezoceramics replaced $BaTiO_3$ ceramics in most applications and PZT remains one of the most popular piezoceramic materials.

PZT is a polycrystalline ferroelectric material. In a ferroelectric material, the internal dipoles of the material can be reoriented by the application of an external electric field, leaving a remnant polarization at zero applied electric field [10]. This remnant polarization also changes with the applied stress and this is how piezoelectricity takes place. Since 1954, there has been

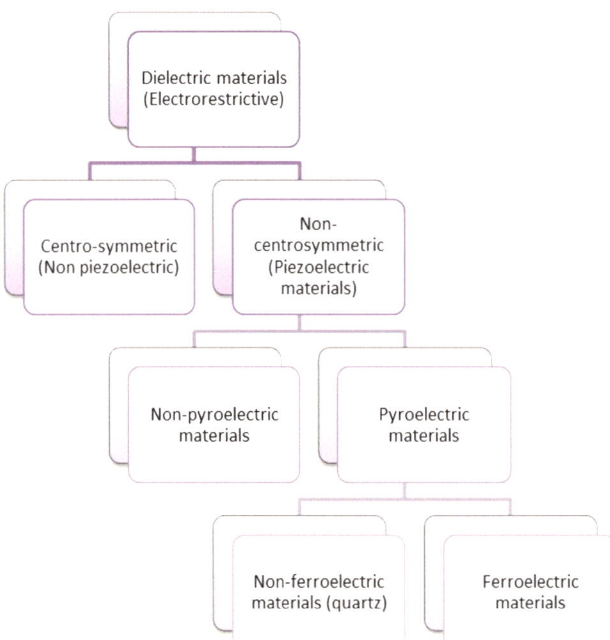

Figure 1. Classification of dielectric materials.

a lot of research to determine the effects of composition (Zr/Ti) and small amounts of additives on the electrical and mechanical properties of PZT piezoceramics [11, 12].

In 1969 Kawai [13] discovered large piezoelectricity in elongated and poled films of polyvinylidene fluoride (PVDF). Research has shown that the polar β-phase of PVDF, which is caused by the application of mechanical stress and/or strong electric fields, is responsible for the development of the piezoelectric property of the material [14, 15].

The piezoelectric behavior of other polar polymers like the odd numbered polyamides such as Polyamide 11 have also been investigated [16–18]. Newman et al. [16] investigated the crystal structure of Polyamide 11, as well as the effect of poling conditions (temperature, time, and poling field), to the overall piezoelectric constants of the material. Polyamide (nylon) is a polymer consisting of the zig-zag chains of CH_2 groups connected by the amide groups (H—N—C=O). The planar sheet structure of molecules is formed by hydrogen bonds between amino groups of adjacent molecules. Scheinbeim et al. [19] used X-ray diffraction to investigate the orientation of the inter-chain hydrogen bonds between the amide bonds, which make up the sheet structure of Polyamide 11, when investigating the polarization of Polyamide 11 in the film form. The planar sheets are oriented parallel to the surface of the film. According to their findings, during poling, the amide dipoles rotate 90° under the strong electric field, which also causes the 90° rotation of the hydrogen-bonded sheets. This rotation results in the 180° rotation of the dipoles.

The molecular structure of odd-numbered (top) and even-numbered (bottom) polyamides is shown in **Figure 2**. In odd-numbered nylons, the electric dipoles formed by amide groups (H—N—C=O) are sequenced in a way that all the dipoles are in the same direction. Therefore, a net dipole moment occurs. In even-numbered nylons, one amide group is in one direction, the next one will be in the opposite direction, alternately. This results in an intrinsic cancelation of the dipole moments [20].

The piezoelectric behavior of polypropylene has mostly been investigated in the case of cellular polypropylene films [21–26], where piezoelectric behavior is a result of the morphology of the structure (air or other gas-filled voids, void morphology, and charge distribution). Other research on the piezoelectric properties of single-film polypropylene or melt-spun polypropylene fibers has been scarce. Research carried out by Kravtsov et al. [27] investigated the polarization of melt-spun polypropylene fibers and concluded that melt-spinning technology favors the formation of spontaneous electret charge in the fibers and that forced fiber polarization in external electric fields gives rise to strong electret effects.

Moreover in the same paper Kravtsov et al. attributed the total electret effect in polypropylene fibers to mechanisms such as Maxwell–Wagner polarization, dipole orientation, and charge carrier injection. Furthermore in 2015, Klimiec et al. [28] investigated the effect of the introduction of SiO_2 and Kaolin fillers on the piezoelectric constant and thermal durability of polypropylene electret, by creating a cellular structure in a single-layer film.

In the research quoted above most of the polymeric piezoelectric materials under investigation were in the film form; however, it is now possible to use polymeric piezoelectric filaments [29, 30] for the applications where flexibility is required, for example, in the interest of producing innovative, "smart" textile products whose components can be integrated into existing textile structures [31–33].

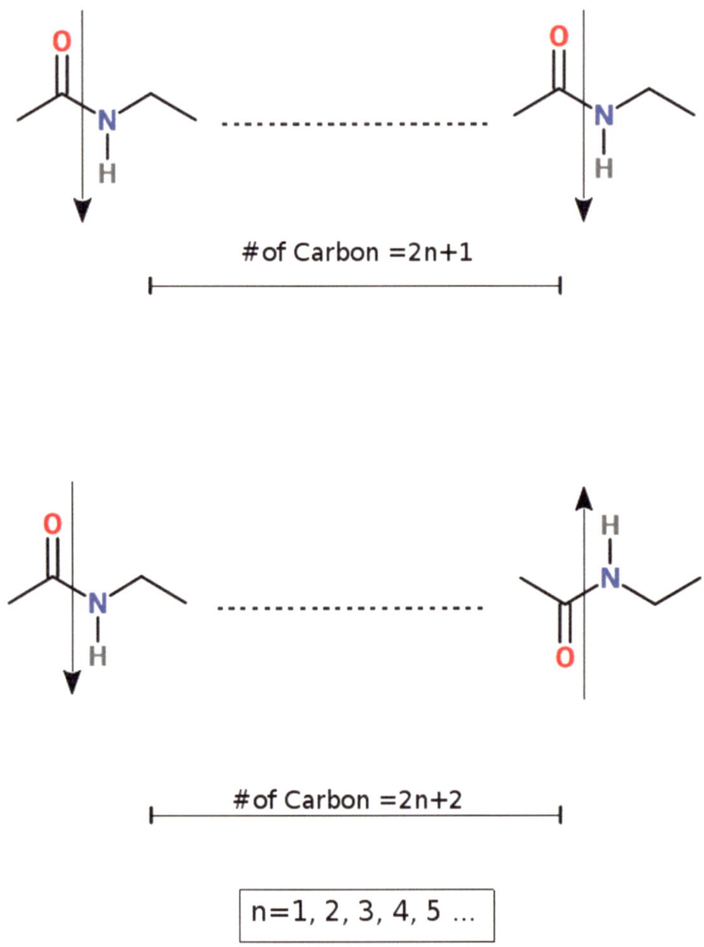

Figure 2. Schematics of molecular structure of odd-numbered and even-numbered polyamides.

While the use of piezoceramic materials such as PZT is extensive, piezoceramics are extremely brittle. Lee et al. [34] compared a PVDF film coated with poly(3,4-ethylenedioxy-thiophene)/poly(4-styrenesulfonate) [PEDOT/PSS] electrodes to films coated with the inorganic electrode materials, indium tin oxide (ITO), and platinum (Pt). When subjected to vibrations of the same magnitude over varying frequencies, it was found that the films with the inorganic-coated electrodes begun to show fatigue cracks at an early stage and at relatively lower frequencies than the PEDOT/PSS film. In further research by Lee et al. [35], piezoceramics tested were susceptible to fatigue crack growth when subjected to high-frequency cyclic loading.

Moreover, while ceramics have a higher piezoelectric constant, the polymers are more flexible making them more appropriate for areas such as wearable applications [36]. Wearable applications, smart textiles, and e-textiles in general (multifunctional textile products) place specific limitations regarding the rigidity, elasticity, thickness, wearability, comfort, and so on of the usually fibrous materials to be incorporated in the product, hence the need for piezoelectric material forms that emulate classic textile structures (fibers, yarns, and fabrics). Multifunctional textile materials become increasingly important for combined applications. Piezoelectric fibers and yarns open a new field in the multifunctional textile area, especially for energy-harvesting applications. It is expected that soon a garment using piezoelectric fibers will be developed capable of producing usable electrical power [37].

2. Melt-spun textile fiber materials as piezoelectric elements

In order to obtain usable textile filaments (filaments are a synonym of the word fiber and are specific to continuous fibers vs. staple fibers) from polymers such as PVDF, the polymer must go through a process known as spinning, that is, the transformation (ordering) of the material into yarn. There are several spinning methods applied to polymers that are already in use in the textiles sector. All methods consist of transforming a solution of the polymer, either produced directly from raw materials (direct spinning) or from dissolving/melting the polymer chips (dry spinning, melt spinning and electrospinning).

Both melt spinning and electrospinning can be utilized to produce piezoelectric polymer filaments. The research presented in this chapter is concerned with filaments produced through melt spinning. In melt spinning, polymer chips are melted and then the melt is forced (extruded) through the spinning head called a spinneret. The holes of the spinneret can have different cross-sectional shapes such as round, trilobal, pentagonal, and so on. Each of the cross-sectional shapes has its own advantages regarding the appearance or properties of the filaments produced. Another available fiber structure is the production of bicomponent filaments. Most of the papers analyzed below are concerned with bicomponent filaments. After production, the filaments are drawn and wound unto bobbins. The drawing (elongation) results in the orientation of the macromolecules of the polymer and improves fiber characteristics such as tensile strength. [38, 39].

The process of producing piezoelectric melt-spun textile fibers as described mainly for PVDF includes one more stage after the final drawing stage used during production of the melt-spun fiber. That stage, poling, is a combination of extension, heating, and exposure to high voltage. Extension of the polymer structure (drawing) together with an elevated temperature allows for the transformation from the α-phase crystallites to β-phase. Then, to orient the dipole moments of the β-phase crystallites (rendering the structure as polar), PVDF is subjected to a high electric field. In the specific case of PVDF the stretch ratio and the temperature at which poling is realized affect the maximum β-phase content, which is as previously discussed directly responsible for the development of the piezoelectric property of PVDF [5]. Typical conditions of poling are 80–90°C and the drawing ratio of 5:1 [40–43].

Figure 3 displays the continuous method for the production of melt-spun piezoelectric textile fibers developed at the University of Bolton. Melt extrusion of the fibers is carried out using a single screw laboratory line melt extruder (Plasticisers Engineering, UK). The extruder screw (diameter of 22 mm) can be operated at speeds up to 50 rpm. The actual speed used for feeding the polymer through the screw is 2 rpm. A flat temperature profile is used for all the polymers consisting of a hopper temperature of 190°C, with a 10°C increment along the barrel and a temperature of 230°C set at the die head.

The apparatus has two water-cooled take-up slow rollers, four temperature-controlled slow rollers, and two fast rollers. The water-cooled rollers are used for the additional cooling of the fiber and temperature-controlled rollers heated the fiber to the required poling temperature, 80°C in this case. The space between temperature-controlled slow rollers and the fast rollers housed a pair of flat-plate electrodes separated by a gap of 10 mm.

To produce the electric field, a Spellman SL300 series high-voltage power supply with a range of 0–20 kV at an output current of 3 mA is used. The poling temperature (80°C) is maintained during the polarization step by heating the bottom electrode. There is a speed difference between the fast rollers and the slow rollers where the speed of the fast rollers is five times higher than the speed of the slower ones, thus obtaining a draw ratio of 5:1. At this point a high voltage of 13 kV is applied. The poling conditions (temperature, extension, and high voltage) are applied simultaneously on the fibers between the temperature-controlled slow rollers and fast rollers [44].

Reviewing the results of the search of the literature, regarding melt-spun textile piezoelectric fibers, it became evident that certain research papers could be considered part of a continuing study into this subject by a specific research team as well as following a specific theme. The research into the development of a piezoelectric melt-spun textile fiber attempts to utilize the accumulated knowledge on the piezoelectricity of thin polymer films and mainly thin films made of PVDF. This explains the concentration of the literature on fibers made of PVDF with very few exceptions, which will be discussed below.

One of the challenges that researchers into piezoelectric textile fibers are faced with is the difficulty of indicating and maintaining the orientation of the fibers with regard to the polarization process. The cross-section shape of the fibers produced is the typical circular cross-section used for the manufacture of synthetic textile fibers. Since the polarization of the fibers is carried out across the thickness of the fiber, the orientation of the charges is along opposite ends of the diameter of the fiber. In a freely twisting and turning fiber, it is quite difficult to

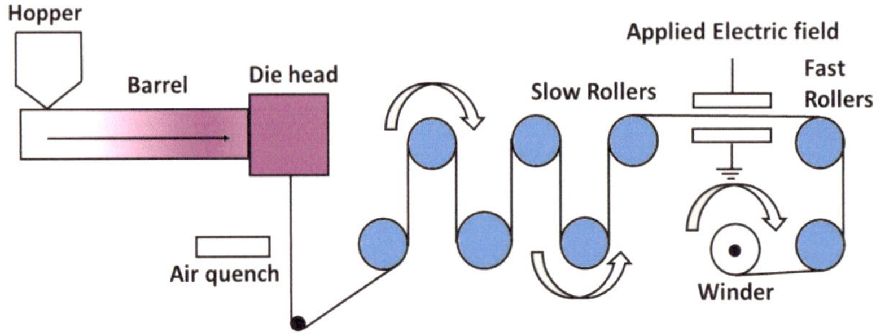

Figure 3. A schematic of the continuous method of production of piezoelectric melt-spun fibers.

determine the position and to orient the theoretical positive (+) side and negative (−) sides of the fiber. Furthermore, the polymer itself, that is, the fiber, has infinite theoretical resistance; this means that the propagation of charges within a complicated structure will be problematic, hence the interest in core-spun fibers with a piezoelectric sheath and a conductive core. A case to point, that will be presented in the following pages, is the work done in France on the production of a piezoelectric coaxial filament, which had a sheath of P(VDF70-TrFE30) (poly(vinylidenefluoride-trifluoroethylene)) and a copper monofilament yarn as core.

As discussed below, the published research is centered mainly on the definition of the parameters for the production of piezoelectric fibers and less on the demonstration of the actual efficacy of the fibers for actual applications (sensors/energy harvesting) or the behavior regarding the aging of the fibers.

Due to the number of researchers working in each team the presentation of the papers that were studied will be geographically grouped (location of the organizations involved in the research).

In Sweden research into melt-spun piezoelectric textile fibers has resulted in a patent [45] and a number of research papers. The first paper, chronologically, by Lund and Hagström [46], investigated the influence of spinning parameters on the β-phase crystallinity of PVDF yarns with no additives or conductive cores. Beginning with the next paper again by Lund and Hagström [47] the researchers introduced the concept of bicomponent PVDF fibers (i.e., PVDF filaments with a conductive core). The conductive cores used in the research originating in Sweden included electrically conductive composites of carbon black (CB) and high-density polyethylene (HDPE) [47, 48], either non-functionalized or amino-functionalized double-wall carbon nanotubes (DWNT) [49] and ethylene-octane copolymer and CB or a high-density polyethylene and again CB [50]. Guo et al. [51] carried out a comparison between three compositions of piezoelectric fibers based on PVDF, that is, PVDF fibers, PVDF/ nanoclay fibers and PVDF/NH2-DWCNT (amino-modified double-wall carbon nanotubes). Finally, in a paper by Nilsson et al. [52] the composition of fibers under investigation is given as a bicomponent with a PVDF sheath and a conductive core while pointing at a previous paper [47] for more information.

Concerning the methods used for the characterization of the fibers (filaments) examined in the papers mentioned above, these are as follows (**Table 1**).

Regarding the research originating in the UK, in 2011, Vatansever et al. [33] published a chapter in the book "Smart Woven Fabrics in Renewable Energy Generation" and the chapter included a presentation of the production method for PVDF piezoelectric monofilament yarns. In 2012, Vatansever et al. [36] presented the production process of a PA-11 (Polyamide 11) piezoelectric monofilament yarn. Vatansever et al. [53] and Hadimani et al. [29] investigated the properties of a PVDF monofilament yarn. In 2015, Bayramol et al. [54] investigated the effect of the addition of multiwalled carbon nanotubes on the piezoelectric properties of polypropylene filaments.

Further on the research carried out in the UK, there was a joint research published that was carried out by researchers based in the UK and researchers based in Greece. The research

Method	Research papers
Differential scanning calorimetry (DSC)	[46–48]
X-ray diffraction (XRD)	[46, 47, 49, 50]
Determination of tensile strength to break	[46, 48, 50]*
Determination of viscosity as a function of shear rate	[50]
Electrical (DC) conductivity measurements	[48, 50]
Determination of the resistance and capacitance of the sensor (individual filament lengths oriented in parallel) and electromechanical characterization of the sensor by subjecting it to a dynamic compression strain perpendicular to the fiber axis	[50]
Determination of the electric signal and strain of a yarn comprising of 24 fibers by subjecting it to a dynamic tensile strain parallel to the fiber/ yarn axis and estimations of the mean power from the fibers	[48]
Evaluation of the sensor (yarn woven into fabric) properties for heartbeat detection	[48]
Characterization of the piezoelectric fibers by connecting the fiber to an impedance analyzer	[52]

*Testing carried out in yarn form.

Table 1. Characterization methods used in research papers originating in Sweden.

was carried out on piezoelectric melt-spun fibers that were produced based on the process developed by Siores et al. [30] with a composition of PVDF, PP and PA-11 with two different cross-sections (ribbon yarns and cylindrical monofilaments). Matsouka et al. published research concerned with the durability of the electrical response (peak-to-peak voltage) of the fibers after one wash cycle [44] as well as a paper describing a method/device that could be used to measure the electrical power produced by the fibers [55].

The UK-based research/joint research constitutes the only research that investigated materials other than PVDF, namely PA-11 [36, 44, 55] and Polypropylene [44, 54, 55]. They also hold the oldest patent [30] on the production of piezoelectric melt-spun textile filaments, the process described in detail by Hadimani et al. [29]. Vatansever et al. [36] touch on the subject of the amount of energy produced by a single filament versus the energy required for powering small electronic devices, though without providing specific data regarding the energy produced. In contrast, Matsouka et al. [55] provide power measurement results for the three different compositions and two different cross-sections examined.

A significant by-product of the research by Matsouka et al. [44] was a conference paper by Vossou et al. [56] who combined the electrical response (peak-to-peak voltage) produced by the piezoelectric fibers with a computational investigation of the mechanical behavior of the same piezoelectric fibers. Piezoelectric fibers were the subject of modal analysis with the use of the finite elements method to evaluate its eigenfrequencies and mode shapes (modal analysis is the study of the dynamic properties of systems in the frequency domain, a typical example would be testing structures under vibrational excitation).

http://dx.doi.org/10.5772/intechopen.78389

Furthermore, by comparing the diagram produced by plotting the bending, y-axis, reaction moment developed at the clamped end of the fibers versus time, to the diagram of the deflection of the free end of the fibers, it was found that in the diagram of the bending, y-axis, reaction moment resembled strongly the typical waveform produced during periodic stimulation of piezoelectric ribbon fibers when plotting the voltage versus time. These findings suggested that the production of electric power through the stimulation of the fibers is confined to the clamped area of the fiber, that is, the specific area of the fiber that is being bended (**Table 2**).

In Germany, in 2010, Walter et al. [57] manufactured melt-spun PVDF fibers of textile finesse. Apart from the typical production processes, the produced filaments underwent false twist texturizing. In 2011, Steinmann et al. [58], produced melt-spun PVDF textile fibers using different production parameters. Also in 2011, Walter et al. [59] carried out the characterization of composites made by combining piezoelectric PVDF monofilaments with a two-composite epoxy resin. In 2012, Walter et al. [60] further developed the previous research project [57] by producing both a warp-knitted fabric and two woven fabrics (plain and twill weave). In 2013, Glauß et al. [61] investigated the spin-ability and characteristics of PVDF bicomponent fibers with a CNT/ PP core. This research project is related to research done by Steinmann et al. [62] on the extrusion of CNT-modified polymers. In 2015 Glauß et al. [63] worked on the production and functionalizing of bicomponent fibers consisting of PVDF "sheath" and conductive CNT/PP cores. Also in 2015 Glauß et al. [64] presented their research in the 4th International Conference on Materials and Applications for Sensors and Transducers. The presentation was regarding the poling effect on bicomponent piezoelectric fibers (PVDF sheath with carbon nanotubes as the core).

The research by Steinmann et al. [58] into the phase transitions of melt-spun PVDF fibers was a significant step in determining the effect of process parameters on the crystallinity of

Method	Research papers
Fourier transform infrared spectroscopy (FTIR)	[29, 44, 55]
X-ray powder diffraction (XRD)	[29]
Differential scanning calorimetry (DSC)	[44, 55]
Determination of linear density	[29, 53]
Determination of tensile strength	[53, 54]
Examination of the micro structures of the filaments under scanning electron microscope (SEM)	[29, 53, 54]
Determination of the electric response in Volts of a group of fibers when stimulated by a mechanical stimulus (impact)	[33, 36, 53, 54]
Determination of the electrical response (peak-to-peak voltage) of a single fiber when stimulated by a mechanical stimulus (bending)	[44]
Determination of the electric power (Watts) produced by a single fiber when stimulated by a mechanical stimulus (bending)	[55]

Table 2. Characterization methods used in research papers originating in the UK.

PVDF. The aim of the research was to understand the crystallization and phase transitions in PVDF fibers in order to optimize the formation of the β phase (which is connected to the piezoelectric behavior of PVDF). The research resulted in a detailed overview of the effect of production properties on the phase transformations of PVDF, which is shown in **Figure 4**.

Walter et al. [59] constructed composite specimens using a sandwich of PVDF monofilaments placed parallel to each other and an epoxy resin, placed between copper films. Polarization was carried out on the composite specimens in an oil bath. Determination of the piezoelectric behavior of the samples was carried out both parallel and perpendicular to the fibers. The specimens were subjected to tensile strain and the voltage produced was measured. The results showed anisotropy of the behavior of the composite specimen regarding the voltage produced depending on the direction of the strain (lengthwise or perpendicular to the length) (**Table 3**).

In Portugal, in 2011, Ferreira et al. [65] investigated the effect of processing conditions and a conductive inner core on the electroactive phase content and the mechanical properties of PVDF filaments without a core and with a core containing a conductive PP (polypropylene)/ carbon black composite. In 2013, Silva et al. [66] investigated the effect of repeated processing cycles on crystallinity and the electroactive phase content of recycled PVDF filaments. In 2014, Martins et al. [67] examined the properties of piezoelectric coaxial filaments. The test specimens comprised a piezoelectric cable obtained from a two-layer coextruded filament, comprising an internal semi-conductive electrode (carbon black-filled polypropylene compound and a carbon nanotube-based compound) and a PVDF layer, coated with a thin layer of a

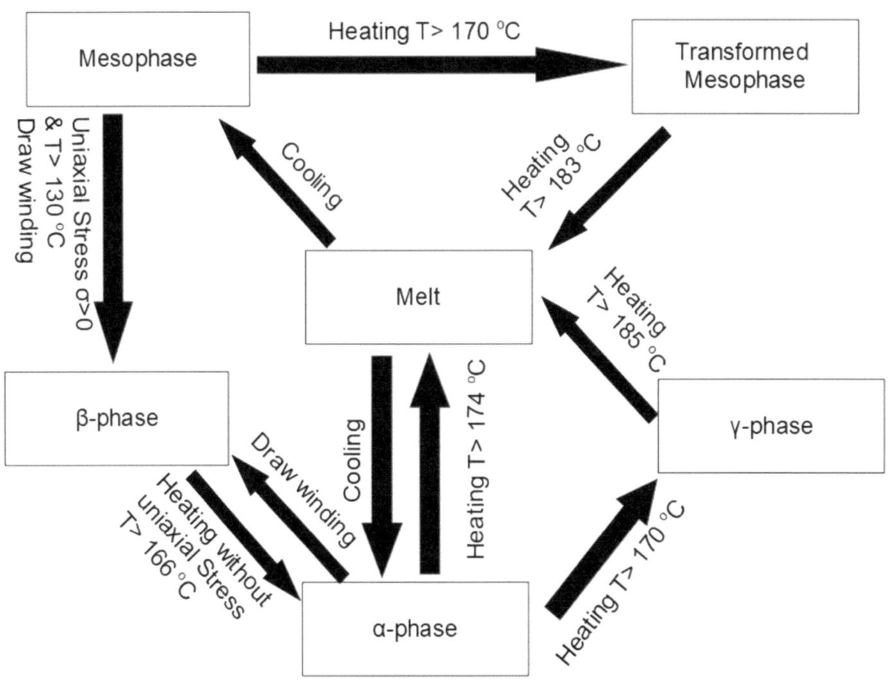

Figure 4. The possible structural phase transitions in fibrous PVDF.

Method	Research papers
Differential scanning calorimetry (DSC)	[57–59, 62]
Wide angle, x-ray diffraction (XRD)	[57–59, 62]
Scanning electron microscope SEM	[57]
Determination of yarn finesse	[57]
Determination of tensile strength at break	[57, 59]
Determination of hot air shrinkage	[57]
Dynamic mechanical analysis (DMA)	[58]
Monitoring of the formation of surface charges on the composites under tensile and bending deformation	[59]
Rheometry measurements	[62]
Transition electron microscopy	[62, 63]
Determination of specific resistivity	[62, 63]
Bright field microscopy	[62]

Table 3. Characterization methods used in research papers originating in Germany.

Method	Research papers
Wide angle, X-ray diffraction (XRD)	[65, 66]
Tensile strength tests to determine the Young modulus of the fibers	[65]
Fourier transform infrared spectroscopy (FTIR)	[66, 67]
Determination of tensile strength at break	[67, 68]
Measurement of the electric conductivity	[67]
Measurement of the electromechanical response (Voltage response of the filaments during mechanical stimulation [tensile strain])	[67]
Microscopy	[67]
Determination of the electromechanical response of the filaments (voltage produced due to mechanical stimulation (vibration, elongation)	[68]

Table 4. Characterization methods used in research papers originating in Portugal.

semi-conductive copper-based lacquer. Also in Portugal, in 2014, Rui et al. [68] investigated coaxial PVDF filaments with a filament core comprising conductive PP. Ferreira et al. [65] concentrated their efforts on producing coaxial piezoelectric filaments made from PVDF, as opposed to pure polymer filaments. The use of a conductive PP/carbon black composite core in the filaments is a common approach in other research, for instance, the work by researchers based in Sweden [48–52] and in Germany [61, 63, 64].

Method	Research papers
Differential scanning calorimetry (DSC)	[70]
Wide angle, x-ray diffraction (XRD)	[70, 71]
Fourier transform infrared spectroscopy (FTIR)	[71]
Determination of tensile strength	[70, 71]
Determination of the sensing capabilities of a woven fabric incorporating the coaxial filaments (voltage response to compression)	[70]
Determination of molecular orientation using optical birefringence	[71]
Determination of the electric response (Voltage) of the piezoelectric fibers integrated in to a woven textile structure after mechanical stimulation (compression)	[71]

Table 5. Characterization methods used in research papers originating in France/joint paper from Australia & Germany.

The subject of the research by Silva et al. [66] (recycled PVDF filaments) was unique in the literature reviewed. The results of several consecutive processing cycles on piezoelectric PVDF samples showed that all the parameters that were studied were unaffected or only very slightly affected by up to nine processing cycles suggesting that PVDF recycling was feasible regarding its electroactive properties (**Table 4**).

In France, Kechiche et al. [69] investigated the properties of a piezoelectric coaxial filament, which had a sheath of P(VDF70-TrFE30) (poly(vinylidenefluoride-trifluoroethylene)) and a copper monofilament as the core. Their work was based on previous research carried out by Khoffi et al. [70] on the production of a polyethylene terephthalate/copper composite filament.

In a joint paper by researchers based in Australia and Germany, Magniez et al. [71] investigated the effect of drawing on the molecular orientation and polymorphism of melt-spun PVDF fibers. The methods used for the characterization of the fibers were (i) the determination of tensile properties of the fibers, (ii) X-ray diffraction (XRD), (iii) Fourier-transform infrared spectroscopy (FTIR), (iv) the determination of molecular orientation using optical birefringence, and (v) the determination of the electric response (voltage) of the piezoelectric fibers integrated in to a woven textile structure after mechanical stimulation (compression).

The approach by Kechiche et al. [69] of manufacturing and studying a coaxial filament (PET/copper) was found to be innovative in the literature reviewed. The research team had to design and develop a new type of spinneret to provide good centering of the inner core (copper filament) in the P(VDF70-TrFE30) matrix copolymer. The research team was able to integrate the monofilament yarns into a woven fabric structure and use the resultant fabric as a pressure sensor (**Table 5**).

3. Conclusions

Based on the analysis of the literature presented in this chapter with regard to the methods of characterization of piezoelectric melt-spun textiles fibers, three main conclusions can be reached: (i) most of the research carried out focuses on PVDF core-spun fibers with very few exceptions. Examples of the exceptions is the research by Bayramol et al. [54] that investigated the piezoelectric behavior of PP and the work by Matsouka et al. [44, 55] which investigated the behavior of PP and PA-11 fibers as well as PVDF ones, (ii) the majority of the current research utilizes test methods such as XRD, differential scanning calorimetry (DSC) and FTIR to characterize piezoelectric fibers, and (iii) there is no standardized method for the determination of the electrical response of the fibers to mechanical stimulation (neither as a method nor as equipment). Methods such as XRD, DSC, and FTIR aim at the characterization of fiber crystallinity and especially in the case of PVDF, the percentage of β phase, which is the source of the piezoelectric properties for PVDF.

There are two major approaches regarding the characterization of the electromechanical response of the fibers: (i) qualitative tests that are intended to show the potential of the fibers, that is, research conducted by Nilsson et al. and Kechiche et al. [52, 69] and (ii) measurement of the voltage produced by the fibers (or multifilament yarns or fabrics incorporating the said yarns) when they are mechanically stimulated either by tensile strain [48, 67, 68], impact [33, 53, 59], compression [50, 70, 71] or bending [44].

From the electrical point of view, these research approaches restrict themselves in the measurement of the generated voltage, that is, the various piezoelectric fibers were characterized by the maximum voltage generated. Furthermore, in most of the research quoted above, the voltage measurement corresponds to the open-circuit voltage which is used as the main performance indication.

However, considering energy-harvesting applications, the open-circuit voltage is not adequate to characterizing the power-generation capabilities of fibers. For this purpose, knowledge of the current production capabilities of the fibers under load is also required. Research that quotes the measurement of the power produced by the melt-spun piezoelectric textile fibers under load with a description of an innovative method/device developed especially for these fibers can be found in the work of Matsouka et al. [55].

Conflict of interest

There is no conflict of interest.

Author details

Dimitroula Matsouka[1]* and Savvas Vassiliadis[2]

*Address all correspondence to: ptosky@gmail.com

1 University of Bolton, Bolton, UK

2 Piraeus University of Applied Sciences, Egaleo, Greece

References

[1] Curie J, Curie P. Développement, par pression, de l'électricité polaire dans les cristaux hémièdres à faces inclinées. Comptes Rendus de l'Académie des Sciences. 1880;**91**:294-295

[2] Jean-Mistral C, Basrour S, Chaillout J-J. Comparison of electroactive polymers for energy scavenging applications. Smart Materials and Structures. 2010;**19**(8):085012

[3] Abramovich H. Intelligent Materials and Structures. Berlin, Boston: De Gruyter; 2016. p. 378. (De Gruyter Graduate)

[4] Whatmore RW. Piezoelectric and pyroelectric materials and their applications. Electronic Materials. 1991. pp. 283-290

[5] Eguchi M. XX. On the permanent electret. Philosophical Magazine Series 6. 1925;**49**(289): 178-192

[6] Fukada E. Piezoelectricity of wood. Journal of the Physical Society of Japan. 1955;**10**(2): 149-154

[7] Fukada E, Yasuda I. On the piezoelectric effect of bone. Journal of the Physical Society of Japan. 1957;**12**(10):1158-1162

[8] Berlincourt D. Piezoelectric ceramics: Characteristics and applications. The Journal of the Acoustical Society of America. 1981;**70**(6):1586-1595

[9] Ramadan KS, Sameoto D, Evoy S. A review of piezoelectric polymers as functional materials for electromechanical transducers. Smart Materials and Structures. 2014;**23**(3):033001

[10] Setter N, Damjanovic D, Eng L, Fox G, Gevorgian S, Hong S, et al. Ferroelectric thin films: Review of materials, properties, and applications. Journal of Applied Physics. 2006;**100**(5):051606

[11] Heywang W, Thomann H. Tailoring of piezoelectric ceramics. Annual Review of Materials Science. 1984;**14**(1):27-47

[12] Jang S-J. Ferroelectrics. In: Kirk-Othmer Encyclopedia of Chemical Technology [Internet]. Hoboken, NJ, USA: John Wiley & Sons, Inc.; 2000. DOI: 10.1002/0471238961.0605181810011407.a01

[13] Kawai H. A new type of piezoelectricity in elongated polymer films. OYOBUTURI. 1969;**38**(12):1133-1138

[14] Davis GT, McKinney JE, Broadhurst MG, Roth SC. Electric-field-induced phase changes in poly(vinylidene fluoride). Journal of Applied Physics. 1978;**49**(10):4998-5002

[15] Sencadas V, Gregorio R, Lanceros-Méndez S. α to β phase transformation and Microestructural changes of PVDF films induced by uniaxial stretch. Journal of Macromolecular Science Part B. 2009;**48**(3):514-525

[16] Newman BA, Chen P, Pae KD, Scheinbeim JI. Piezoelectricity in nylon 11. Journal of Applied Physics. 1980;**51**(10):5161-5164

[17] Newman BA, Kim KG, Scheinbeim JI. Effect of water content on the piezoelectric properties of nylon 11 and nylon 7. Journal of Materials Science. 1990;**25**(3):1779-1783

[18] Takase Y, Lee JW, Scheinbeim JI, Newman BA. High-temperature characteristics of nylon-11 and nylon-7 piezoelectrics. Macromolecules. 1991;**24**(25):6644-6652

[19] Scheinbeim JI, Lee JW, Newman BA. Ferroelectric polarization mechanisms in nylon 11. Macromolecules. 1992;**25**(14):3729-3732

[20] Su J. Ferroelectric polymers for electromechanical functionality. In: Kim KJ, Tadokoro S, editors. Electroactive Polymers for Robotic Applications: Artificial Muscles and Sensors [Internet]. London: Springer London; 2007. pp. 91-120. DOI: 10.1007/978-1-84628-372-7_4

[21] Hillenbrand J, Sessler GM. Quasistatic and dynamic piezoelectric coefficients of polymer foams and polymer film systems. IEEE Transactions on Dielectrics and Electrical Insulation. 2004;**11**(1):72-79

[22] Lekkala J, Poramo R, Nyholm K, Kaikkonen T. EMF force sensor—A flexible and sensitive electret film for physiological applications. Medical & Biological Engineering & Computing. 1996;**34**(Suppl. 1):67-68

[23] Neugschwandtner GS, Schwödiauer R, Vieytes M, Bauer-Gogonea S, Bauer S, Hillenbrand J, et al. Large and broadband piezoelectricity in smart polymer-foam space-charge electrets. Applied Physics Letters. 2000;**77**(23):3827-3829

[24] Paajanen M, Lekkala J, Kirjavainen K. ElectroMechanical Film (EMFi)—A new multipurpose electret material. Sensors and Actuators A: Physical. 2000;**84**(1-2):95-102

[25] Peltonen J, Paajanen M, Lekkala J. Determination of the actuator sensitivity of electromechanical polypropylene films by atomic force microscopy. Journal of Applied Physics. 2000;**88**(8):4789-4793

[26] Sessler GM, Hillenbrand J. Electromechanical response of cellular electret films. Applied Physics Letters. 1999;**75**(21):3405-3407

[27] Kravtsov AG, Brünig H, Zhandarov SF. Analysis of the polarization state of melt-spun polypropylene fibers. Journal of Materials Processing Technology. 2002;**124**(1-2):160-165

[28] Klimiec E, Królikowski B, Machnik M, Zaraska W, Dzwonkowski J. Increase of piezoelectric constant and thermal durability of polypropylene electret by introducing SiO_2 and kaolin filler and creating a cellular structure. Journal of Electronic Materials. 2015;**44**(7):2283-2291

[29] Hadimani RL, Bayramol DV, Sion N, Shah T, Qian L, Shi S, et al. Continuous production of piezoelectric PVDF fibre for e-textile applications. Smart Materials and Structures. 2013;**22**(7):075017

[30] Siores E, Hadimani RL, Vatansever D. Piezoelectric Polymer Element & Production Method & Apparatus Therefor. GB Patent Application 1015399.7. GB 2010

[31] Siores E, Vassiliadis S, Matsouka D. Energy harvesting and wearable electronics from the textile engineer's perspective. In: eRA-6 International Scientific Conference Proceedings, Athens, Greece; 21-24 September 2011

[32] Soin N, Shah TH, Anand SC, Geng J, Pornwannachai W, Mandal P, et al. Novel "3-D spacer" all fibre piezoelectric textiles for energy harvesting applications. Energy & Environmental Science. 2014;**7**(5):1670

[33] Vatansever D, Siores E, Hadimani RL, Shah T. Smart Woven Fabrics In: Renewable Energy Generation, Advances in Modern Woven Fabrics Technology. In: Vassiliadis SG, editor. IntechOpen; July 27th, 2011. DOI: 10.5772/24790. Available from: https://www.intechopen.com/books/advances-in-modern-woven-fabrics-technology/smart-woven-fabrics-in-renewable-energy-generation

[34] Lee CS, Joo J, Han S, Koh SK. Multifunctional transducer using poly (vinylidene fluoride) active layerand highly conducting poly (3,4-ethylenedioxythiophene) electrode: Actuator and generator. Applied Physics Letters. 2004;**85**(10):1841-1843

[35] Lee JB, Subramanian V. Weave patterned organic transistors on fiber for E-textiles. IEEE Transactions on Electron Devices. 2005;**52**(2):269-275

[36] Vatansever D, Hadimani RL, Shah T, Siores E. Piezoelectric mono-filament extrusion for green energy applications from textiles. Journal of Textiles and Engineer Tekstil Ve Mühendis. 2012;**19**(85):1-5

[37] Jost K, Dion G, Gogotsi Y. Textile energy storage in perspective. Journal of Materials Chemistry A. 2014;**2**(28):10776-10787

[38] Cook JG. Man-made fibres. In: Handbook of Textile Fibres. 5th ed. Cambridge: Woodhead Publishing; 2005. p. 723

[39] Mohammadi B, Yousefi AA, Bellah SM. Effect of tensile strain rate and elongation on crystalline structure and piezoelectric properties of PVDF thin films. Polymer Testing. 2007;**26**(1):42-50

[40] Gomes J, Serrado Nunes J, Sencadas V, Lanceros-Mendez S. Influence of the β-phase content and degree of crystallinity on the piezo- and ferroelectric properties of poly(vinylidene fluoride). Smart Materials and Structures. 2010;**19**(6):065010

[41] Murase Y, Nagai A. Melt spinning. In: Advanced Fiber Spinning Technology [Internet]. Cambridge, England: Elsevier; 1994. pp. 25-64. Available from: http://linkinghub.elsevier.com/retrieve/pii/B978185573182050008X

[42] Sajkiewicz P, Wasiak A, Gocłowski Z. Phase transitions during stretching of poly(vinylidene fluoride). European Polymer Journal. 1999;**35**(3):423-429

[43] Sobhani H, Razavi-Nouri M, Yousefi AA. Effect of flow history on poly(vinylidine fluoride) crystalline phase transformation. Journal of Applied Polymer Science. 2007;**104**(1): 89-94

[44] Matsouka D, Vassiliadis S, Vatansever Bayramol D, Soin N, Siores E. Investigation of the durability and stability of piezoelectric textile fibres. Journal of Intelligent Material Systems and Structures. 2017;**28**(5):663-670

[45] Hagström B, LUND A, Nilsson E, Ab SI. Method of Producing a Piezoelectric and Pyroelectric Fiber. WO Patent Application WO2014161920A1; 2014. Available from: https://www.google.com/patents/WO2014161920A1?cl=it

[46] Lund A, Hagström B. Melt spinning of poly(vinylidene fluoride) fibers and the influence of spinning parameters on β-phase crystallinity. Journal of Applied Polymer Science. 2010;**116**(5):2685-2693

[47] Lund A, Hagström B. Melt spinning of β-phase poly(vinylidene fluoride) yarns with and without a conductive core. Journal of Applied Polymer Science. 2011;**120**(2):1080-1089

[48] Nilsson E, Lund A, Jonasson C, Johansson C, Hagström B. Poling and characterization of piezoelectric polymer fibers for use in textile sensors. Sensors and Actuators A: Physical. 2013;**201**:477-486

[49] Lund A, Gustafsson C, Bertilsson H, Rychwalski RW. Enhancement of β phase crystals formation with the use of nanofillers in PVDF films and fibres. Composites Science and Technology. 2011;**71**(2):222-229

[50] Lund A, Jonasson C, Johansson C, Haagensen D, Hagström B. Piezoelectric polymeric bicomponent fibers produced by melt spinning. Journal of Applied Polymer Science. 2012;**126**(2):490-500

[51] Guo Z, Nilsson E, Rigdahl M, Hagström B. Melt spinning of PVDF fibers with enhanced β phase structure. Journal of Applied Polymer Science. 2013;**130**(4):2603-2609

[52] Nilsson E, Mateu L, Spies P, Hagström B. Energy harvesting from piezoelectric textile fibers. Procedia Engineering. 2014;**87**:1569-1572

[53] Vatansever D, Hadimani RL, Shah T, Siores E. Characterisation of energy generating polyvinylidene fluoride (PVDF) based piezoelectric filament. Advances in Materials Research. 2011;**410**:366-369

[54] Bayramol DV, Soin N, Hadimani RL, Shah TH, Siores E. Effect of addition of multiwalled carbon nanotubes on the piezoelectric properties of polypropylene filaments. Journal of Nanoscience and Nanotechnology. 2015;**15**(9):7130-7135

[55] Matsouka D, Vassiliadis S, Prekas K, Bayramol DV, Soin N, Siores E. On the measurement of the electrical power produced by melt spun piezoelectric textile fibres. Journal of Electronic Materials. 2016;**45**(10):5112-5126

[56] Vossou CG, Vassiliadis SG, Matsouka D, Siores E. Computational Modelling of Piezoelectric Fibres. In: International Istanbul. Textile Congress 2013 on Innovative and Functional Textiles Proceedings. Istanbul, Turkey; May 30-June 1, 2013

[57] Walter S, Steinmann W, Gries T, Seide G, Schedukat N, Roth G. Melt-spun polyvinylidene-fluoride fibers of textile fineness. Chemical Fibers International. 2010;**60**(1):49-51

[58] Steinmann W, Walter S, Seide G, Gries T, Roth G, Schubnell M. Structure, properties, and phase transitions of melt-spun poly(vinylidene fluoride) fibers. Journal of Applied Polymer Science. 2011;**120**(1):21-35

[59] Walter S, Steinmann W, Schütte J, Seide G, Gries T, Roth G, et al. Characterisation of piezoelectric PVDF monofilaments. Materials and Technologies. 2011;**26**(3):140-145

[60] Walter S, Steinmann W, Seide G, Gries T, Roth G. Development of innovative fibre materials for technical applications—Fine polyvinylidene fluoride filaments and fabrics. Filtration. 2012;**12**(1):60-64

[61] Glauß B, Steinmann W, Walter S, Beckers M, Seide G, Gries T, et al. Spinnability and characteristics of polyvinylidene fluoride (PVDF)-based bicomponent fibers with a carbon nanotube (CNT) modified polypropylene core for piezoelectric applications. Materials. 2013;**6**(7):2642-2661

[62] Steinmann W, Vad T, Weise B, Wulfhorst J, Seide G, Gries T, et al. Extrusion of CNT-modified polymers with low viscosity—Influence of crystallization and CNT orientation on the electrical properties. Polymers and Polymer Composites. 2013;**21**(8):473-482

[63] Glauß B, Bettermann I, Steinmann W, Walter S, Seide G, Gries T, et al. Production and functionalizing of bicomponent fibers based on PVDF. Chemical Fibers International. 2015;**66**(Yearbook):70-72

[64] Glauß B, Jux M, Walter S, Kubicka M, Seide G, Wierach P, et al. Poling effects in melt-spun PVDF bicomponent fibres. Key Engineering Materials. 2015;**644**:110-114

[65] Ferreira A, Costa P, Carvalho H, Nobrega JM, Sencadas V, Lanceros-Mendez S. Extrusion of poly(vinylidene fluoride) filaments: Effect of the processing conditions and conductive inner core on the electroactive phase content and mechanical properties. Journal of Polymer Research. 2011;**18**(6):1653-1658

[66] Silva MP, Martins RS, Carvalho H, Nobrega JM, Lanceros-Mendez S. Extrusion of poly(vinylidene fluoride) recycled filaments: Effect of the processing cycles on the degree of crystallinity and electroactive phase content. Polymer Testing. 2013;**32**(6):1041-1044

[67] Martins RS, Gonçalves R, Azevedo T, Rocha JG, Nóbrega JM, Carvalho H, et al. Piezoelectric coaxial filaments produced by coextrusion of poly(vinylidene fluoride) and electrically conductive inner and outer layers. Journal of Applied Polymer Science. 2014;**131**(17)

[68] Rui M, Marco S, Renato G, Gerardo R, Miguel NJ, Helder C, et al. Processing and electrical response of fully polymer piezoelectric filaments for E-textiles applications. Journal of Textile Engineering. 2014;**60**(2):27-34

[69] Kechiche MB, Bauer F, Harzallah O, Drean J-Y. Development of piezoelectric coaxial filament sensors P(VDF-TrFE)/copper for textile structure instrumentation. Sensors and Actuators A: Physical. 2013;**204**:122-130

[70] Khoffi F, Khenoussi N, Harzallah O, Drean JY. Mechanical behavior of polyethylene terephthalate/copper composite filament. Physics Procedia. 2011;**21**:240-245

[71] Magniez K, Krajewski A, Neuenhofer M, Helmer R. Effect of drawing on the molecular orientation and polymorphism of melt-spun polyvinylidene fluoride fibers: Toward the development of piezoelectric force sensors. Journal of Applied Polymer Science. 2013;**129**(5, 5):2699-2706

Chapter 5

Piezoelectric Vibration Energy Harvester Using Polyvinylidene Difluoride Film Formed by Bar-Coating Method and Its Spray-Coating Method on a Three Dimensional Surface

Hiroki Takise, Masato Suzuki,
Tomokazu Takahashi and Seiji Aoyagi

Additional information is available at the end of the chapter

http://dx.doi.org/10.5772/intechopen.79192

Abstract

A cantilever-type vibration energy harvester (VEH) made of polyvinylidene difluoride (PVDF) was fabricated and characterized. PVDF is one of the polymer piezoelectric materials, which is more flexible than ceramic-based piezoelectric materials such as lead zirconate titanate (PZT). The fabrication process of VEH is as follows: a PVDF film was coated on a phosphor bronze plate by bar-coating method, followed by polarization by corona discharge method. Aluminum top electrode was deposited on the PVDF film by sputtering. One end of the plate was clamped by a fixture to form a cantilever, the length of which is 25 mm. Output power P at the resonance frequency (=55 Hz) was measured as a function of load resistance R with the acceleration set at 17 m/s^2. Maximum output reached 4.3 μW at R = 2.1 MΩ. This result is not inferior compared with other reported VEHs using ceramic piezoelectric material. Spray coating was carried out to form PVDF film on a 3D surface. This method is suitable for fabricating a uniform thin film on a three-dimensional (3D) surface, even if it is complicatedly curved. In this study, PVDF film was formed on a 3D helical compressing spring, and the deposition on it was successfully achieved.

Keywords: piezoelectric polymer, PVDF, thin film, vibration energy harvester, bar coating, spray coating

1. Introduction

Recently, energy harvesting technologies have been actively studied as an energy source for wireless sensor's battery [1, 2]. There are various energy sources for energy harvesting such as solar power, thermal energy, wind power, and vibration. Among these power generation methods, a vibration energy harvester (VEH) is necessary in environments where solar cells cannot be used, for example, bridges, cars, underground places, buildings, and other environments. There are three main types of VEH generation methods: electromagnetic, electrostatic, and piezoelectric. Piezoelectric VEH has high electromechanical coupling coefficient. In addition, its structure is simple and suitable for miniaturization [3]. In many cases, the structure of piezoelectric VEH is like this: piezoelectric thin film is deposited on the surface of leaf spring made of metal or single crystal silicon. Weight is set on the tip of the spring [4–13]. By matching the design of leaf spring to the vibration condition, it can generate electric power under low frequency vibration and low acceleration.

In previous study, Refs. [6–10] fabricated piezoelectric VEHs using ceramic piezoelectric material. They fabricated cantilever-type piezoelectric VEHs. These piezoelectric VEHs were composed of metal cantilever and PZT film having high piezoelectricity on it. The output power of 1–17 μW was reported in the vibration frequency range of several hundred hertz. Although PZT has high piezoelectricity, it contains poisonous lead and it is a typical brittle material. Destruction of the VEH devices under large vibration or impact is a practical problem. In place of PZT, lead-based piezoelectric materials other than PZT and lead-free ones are being studied for VEHs, but piezoelectricity still remains lower than PZT [11, 12].

We adopted polyvinylidene difluoride (PVDF) which is a piezoelectric polymer material. PVDF has greater flexibility compared to ceramic PZT, so it can withstand large deformation. PVDF can be dissolved in an organic solvent such as methyl ethyl ketone (MEK), which forms PVDF solution. It can be coated on substrates or parts by several coating methods, making it compatible with micro electro mechanical systems (MEMS) technology. By using PVDF, therefore, it is possible to fabricate small devices. Toprak et al. fabricated a small piezoelectric VEH device in which a cantilever was made of single crystal silicon coated with a PVDF thin film [13]. However, its vibration frequency was 1074 Hz, and the output power was very small pW order. A VEH, which is coated with PVDF thin film and suitable for low frequency of 100 Hz or lower, has not been reported yet.

At the beginning of this chapter, we aimed to fabricate VEH coated with bar-coated PVDF film on leaf springs which can generate electric power even at low frequency. First, a PVDF thin film was prepared by bar-coating method, and then, it was polarized by corona discharge method. Mechanical and electrical properties of bar-coated films, such as surface roughness, residual stress, Young's modulus, XRD pattern, piezoelectric constant, were characterized. Then, a cantilever-type VEH using a bar-coated PVDF film was fabricated, and its power generating ability was evaluated.

Bar-coating methods enable PVDF film to coat on a completely flat substrate; however, it is not suitable for coating thin film on a three-dimensional (3D) surface (e.g., curved surface, uneven surface) caused by its coating principles.

In contrast, spray-coating method is suitable for coating a thin film on 3D surface. Spray coating was well researched by Sasaki et al. [14]. They have developed spray-coating technique

http://dx.doi.org/10.5772/intechopen.79192

of photoresist onto 3D microstructures for the purpose of photolithography. In the case of PVDF, a pyroelectric image sensor was fabricated using its thin film formed by electrospray coating [15]; however, the film was deposited on only flat substrate. Coating PVDF thin film on completely 3D microstructure has not been reported yet. Recent years, ultramicro 3D fabrication technique such as 3D printing is progressing. A technology to form PVDF thin film on the surface of fine 3D structure is expected. In this chapter, we tried to apply the spray-coating method to PVDF coating on 3D structure.

In the latter part of this study, spray coating was carried out to form PVDF thin film on the surface of 3D helical spring. A state of a spray-coated film was observed by a scanning electron microscope (SEM). The SEM image showed that the spray-coated PVDF thin film deposition was successfully achieved.

2. Preparation of PVDF film by bar-coating method

2.1. Formation of PVDF film

Spin-coating method is typically employed in MEMS process. However, it was found that the thickness of PVDF film can be controlled only in the range of 5–10 μm by this method. Perhaps because of cracks, spin-coated films with a thickness of 5–10 μm were easily discharged in the polarization process. In order to prevent cracks, it was necessary to obtain a thicker film. As a method for obtaining a thick film, bar-coating method was used.

Figure 1 shows a schematic diagram of the bar-coating method. First, drop the PVDF solution onto substrate and slide the bar coater on spacer to expand the PVDF solution. On the surface of the bar-coater, grooves are provided to obtain a uniform film thickness.

PVDF solution was prepared by dissolving PVDF in methyl ethyl ketone (MEK). A bar coater (Model: 065-2-40 μm, Allgood Corp.) having diameters, lengths, and groove depths of 9 mm,

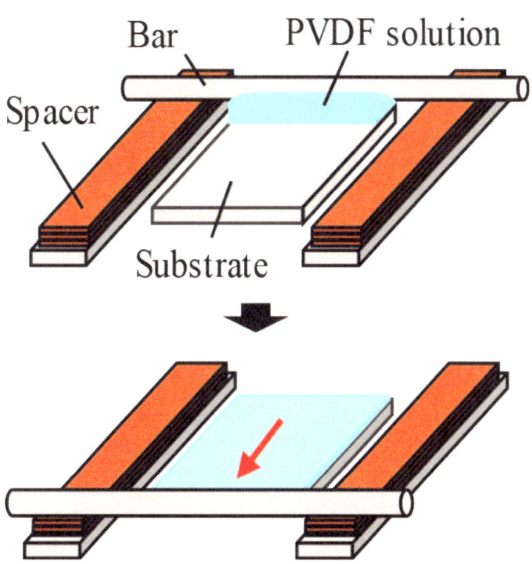

Figure 1. Schematic illustration of bar-coating method.

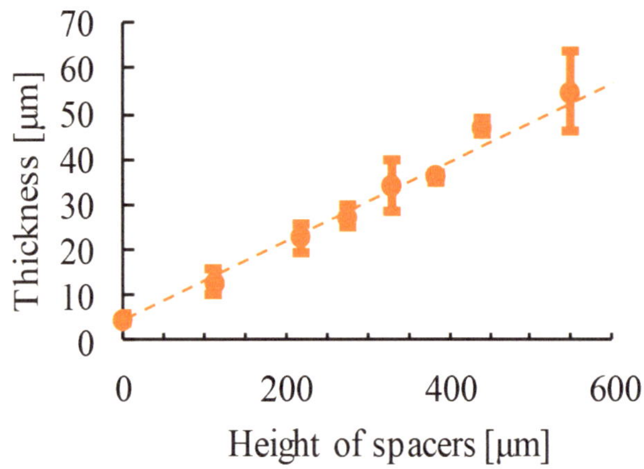

Figure 2. Relationship between obtained thickness of bar-coating PVDF film and spacer height.

320 mm, and 40 μm was used in this study. After coating, the substrate was heated on a hot plate at 90°C for 10 min to evaporate the MEK.

In this method, it was possible to control the film thickness to 50 μm by changing the height of spacer while sliding the bar coater. As a result of the measurement, it was found that the thickness of the bar-coated PVDF film linearly increases in proportion to the height of spacer (see **Figure 2**). However, the surface irregularities increased in the case of the film thickness exceeded 30 μm. Considering this, the thickness was decided to set to 30 μm. Namely, the reason for determining the thickness largely depends on the constraints of the film-forming process. In this chapter, bar-coated films with thicknesses of 30 and 12 μm are prepared and used.

2.2. Characterization of properties of bar-coated PVDF film

Figure 3 shows the surface morphology and roughness of obtained PVDF film measured by a stylus-type step profiler (DekTakXT, Bruker Daltonics K.K). The surface waviness was within ±1.5 μm along 25 mm line (see **Figure 3(a)**). The surface roughness (*R*a) was estimated as

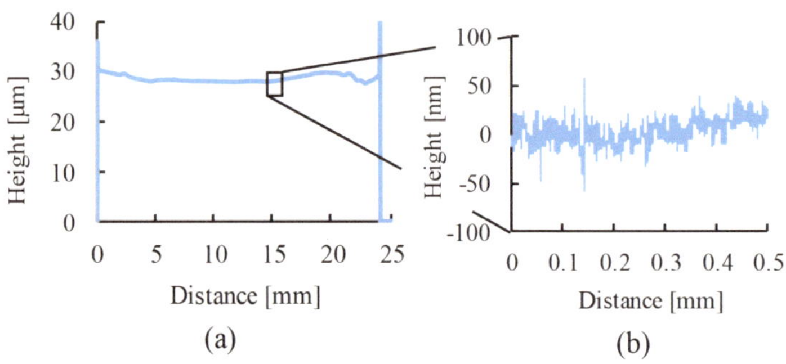

Figure 3. Measurement result by a stylus-type step profiler; (a) surface morphology and (b) surface roughness of PVDF film obtained by bar-coating method.

11 nm using the data shown in **Figure 3(b)**. Observation of the film surface using a scanning electron microscope (SEM) showed no cracks or pinholes of several micrometers.

Residual stress of the PVDF film was measured using a thin-film stress measurement apparatus (FLX–2320–S, Toho technology Corp.). The curvature radius before and after PVDF film formation on a 4-in silicon wafer was measured. A laser beam was scanned along a straight line on the wafer surface; the reflected light was detected to yield a profile of the surface. Using the difference between these radii, the residual stress was calculated as 5 MPa, which is quite low stress compared with usual polymer films [16].

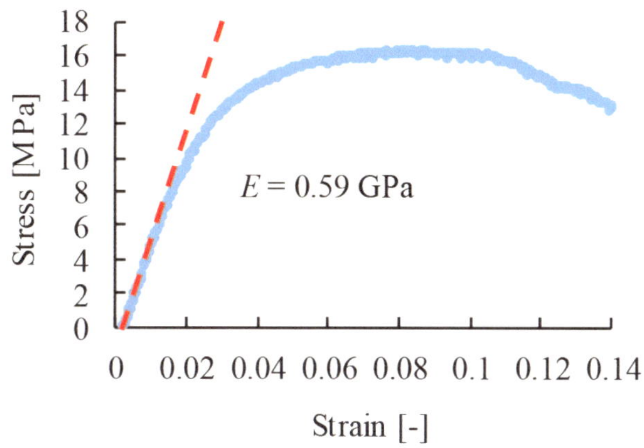

Figure 4. Stress-strain curve of bar-coated PVDF film.

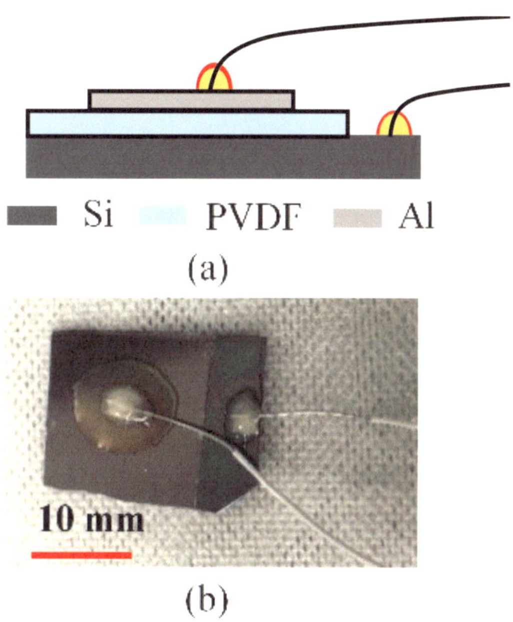

Figure 5. (a) Schematic illustration and (b) photograph of the test sample to measure *P-E* hysteresis loops.

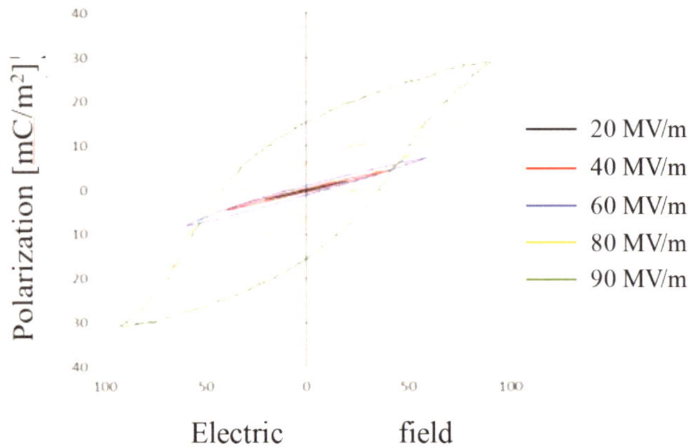

Figure 6. Obtained *P-E* hysteresis loops.

Young's modulus was measured by pulling it physically. The resulting stress-strain curve is shown in **Figure 4**. It has been reported that Young's modulus of bulk PVDF is 1–3 GPa, and that of thin PVDF film is about 0.5 GPa [17, 18]. Therefore, the result of 0.6 GPa calculated by the initial inclination of the curve is considered to be a reasonable value.

Polarization vs. electric field (*P-E*) hysteresis loops for the PVDF film was measured using a Sawyer-Tower circuit. A fabricated test sample was shown in **Figure 5**. PVDF film was formed on a low-resistance silicon substrate (resistivity: 0.2 Ωcm or less) by bar-coating method. Then, aluminum electrode was deposited on PVDF by sputtering, and a lead wire was connected to the electrodes. Electric field was applied by changing it from 20 to 90 MV/m. Obtained *P-E* hysteresis loops were shown in **Figure 6**, which shows good ferroelectricity.

3. Polarization of PVDF film

3.1. Corona polarization method

Following coating a PVDF film, it was polarized by applying high voltage. In early stage of this study, top and bottom electrodes were deposited on PVDF film, then high DC voltage was applied between them for polarizing the film; however, high electric field could not be applied, since discharge often occurred. In almost all cases, the discharge was creeping one, i.e., the route of discharge is on the substrate surface, followed by the route along its side wall to the electrical ground. It indicates that the film itself endures the applied high voltage, i.e., the discharge across the film thickness due to cracks or pinholes does not occur.

In order to solve the discharge problem, charge injection was performed into the PVDF film surface using corona discharge method. In this method, PVDF film was applied high electric field before depositing the top electrode. We call this method as corona polarization. **Figure 7** shows schematic diagram of corona polarization setup. The procedure of this polarization method is as follows: a needle electrode was placed over the center of PVDF film, followed by applying a voltage of −10.0 kV. Distance between the electrode and the test sample was set at 20 mm. Applying time was set for 10 min. In corona polarization, no discharge or any other problems occurred.

http://dx.doi.org/10.5772/intechopen.79192

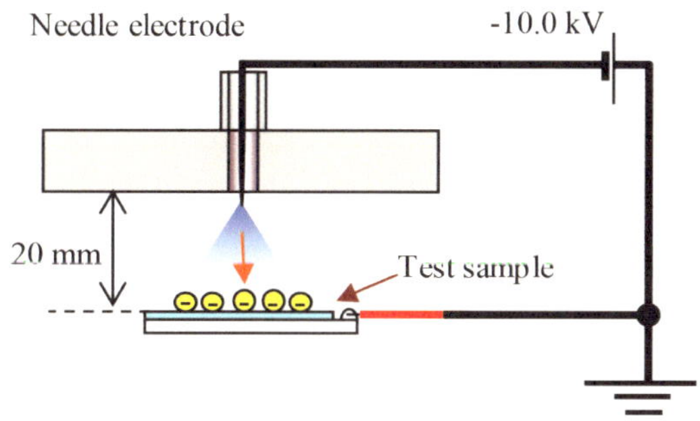

Figure 7. Schematic of corona polarization.

3.2. Characterization of piezoelectric properties

To obtain the information of crystal structure of PVDF film, X-ray diffraction (XRD) analysis was performed. Result of XRD pattern is shown in **Figure 8**. The sharp and narrow peak at 20° indicates the dominant β-phase in PVDF thin film, which is beneficial for the piezoelectric effect. The measured XRD pattern matched well with that of a typical β-phase PVDF film discussed in previous studies [19, 20]. This result shows that bar-coated film had a good crystalline structure.

In order to confirm whether or not the polarized PVDF film has piezoelectricity, the piezoelectric constant d_{33} was measured. The process of fabricating test sample for measuring d_{33} is as follows: As already described in Section 2.1, a PVDF film was prepared on a low-resistance silicon substrate (size: 40 × 30 mm) by bar-coating method. One end of PVDF film was cut, and the lead wire was connected to the exposed low-resistance silicon substrate by a conductive adhesive. The effective area of PVDF film after cutting was 30 mm^2.

Piezoelectric constant d_{33} of PVDF was measured by an apparatus (product name; d_{33} meter, PIEOTEST Corp.). The center of PVDF film right under the needle electrode is a region where charge injection is easy. On the other hand, the amount of charge decreases as going away from the center of film. In other words, the internal electric field generated in PVDF film depends on the location, and there is a possibility that the PVDF film cannot polarize away from the electrode. In order to investigate the success range of one corona polarization, we

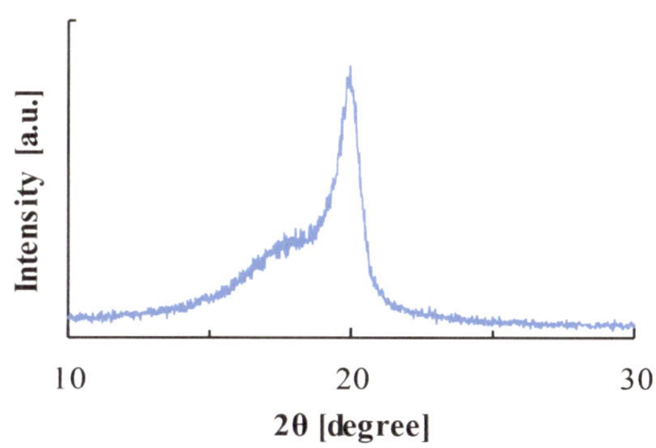

Figure 8. Result of XRD analysis of bar-coated PVDF.

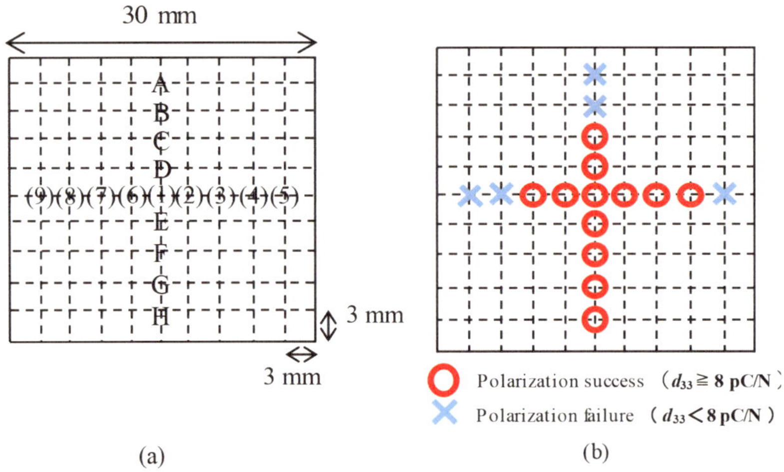

Figure 9. (a) Measurement points of piezoelectric constant d_{33} and (b) result of success or failure of polarization.

	d_{33} [pC/N]		d_{33} [pC/N]
(1)	-9.93	A	-5.59
(2)	-8.86	B	-7.72
(3)	-10.7	C	-8.01
(4)	-9.81	D	-9.46
(5)	-6.89	E	-10.6
(6)	-10.0	F	-11.2
(7)	-8.85	G	-11.1
(8)	-3.40	H	-8.02
(9)	-3.43		

Table 1. Results of piezoelectric constant d_{33} of the PVDF after corona polarization.

measured d_{33} at 17 points at 3 mm intervals. **Figure 9(a)** shows the distribution of measurement points on PVDF film. About 17 points ((1)–(9) and A–H) were adopted as measuring points. **Table 1** shows measurement result of d_{33}. The maximum value of d_{33} was −11.2 pC/N (see **Table 1**). **Figure 9(b)** shows success or failure of corona polarization at each point. When the value of d_{33} was greater than or equals to 8 pC/N, we considered that polarization was successful, otherwise, we considered that polarization was failed. Success range of one corona polarization was within the circle, of which diameter was approximately 15 mm. In the outside area, sufficient piezoelectricity was not achieved.

The resistivity of low-resistance silicon is higher than typical metals, and it may affect the piezoelectricity of PVDF films. We verified the effects of material difference in bottom electrode on piezoelectricity. The d_{33} of PVDF film of 30 and 12 μm in thickness coated on phosphor bronze (2–6 × 10^{-6} Ωcm [21]) was −8.5 pC/N and −10.1 pC/N (see Section 4.1 below), while that was approximately −10 pC/N (distributing among sampling points) on a low-resistance single crystal Si (0.2 Ωcm or less).

Considering these results, although the data is limited, the material of substrate, e.g., the resistivity of substrate, does not affect so much on the piezoelectricity of PVDF film.

4. Fabrication of piezoelectric cantilever-type VEHs

4.1. Preparation of PVDF film for VEHs

PVDF film was formed on a rectangular phosphor bronze plate by bar-coating method. The size of phosphor bronze plate was 35 mm in length, 15 mm in width, and 0.1 mm in thickness. Phosphorus bronze was adopted due to its good spring characteristics. We fabricated two VEHs with different PVDF thicknesses; one is 30 μm (called sample 1) and the other is 12 μm (called sample 2). The dimensions of these samples are shown in **Table 2**.

After forming PVDF film, a lead wire was connected to phosphor bronze plate to perform corona polarization. The conditions of corona polarization were same as described in Section 3.1. In order to securely polarize the entire surface of the PVDF film, corona discharges were performed at six points at intervals of 5 mm (see **Figure 10**).

PVDF thickness sample1	30 μm
PVDF thickness sample2	12 μm
Phosphor bronze thickness	100 μm
Cantilever length	25 mm
Cantilever width	15 mm
Electrode length	22 mm
Electrode width	10 mm

Table 2. Dimensions of cantilever VEH device.

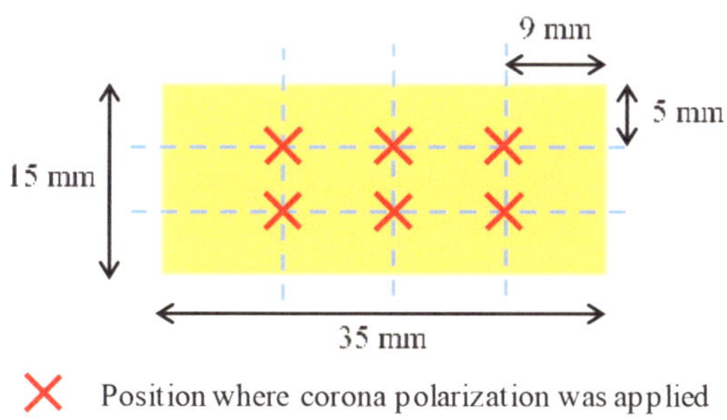

Figure 10. Schematic of points where corona polarization was applied.

After polarization, d_{33} of these PVDF films was measured. The average measured values of d_{33} of polarized PVDF thin films were −8.5 pC/N (sample 1) and −10.1 pC/N (sample 2), respectively. For comparison, d_{33} of PVDF, which was not treated with polarization, was −0.87 pC/N. These values (−8.5 pC/N and −10.1 pC/N) are not bad compared with the typically reported values (−10 to −20 pC/N) [22]. The difference of d_{33} between −8.5 and −10.1 pC/N may be caused by the thickness difference. However, considering only the difference in electric field due to thickness, the difference would be large. It is difficult to identify the reason at present.

In this study, PVDF film operates in d_{31} mode. It is reported that d_{31} of PVDF film is approximately 70% of its d_{33} [23, 24]. Considering that d_{33} of our PVDF film is approximately 10 and it is the same order as other reported values, d_{31} is not bad.

An aluminum upper electrode was deposited on PVDF film by sputtering. The size of electrode was 30 mm in length, 10 mm in width, 0.3 μm in thickness. A capacitance of the prepared VEHs was measured with an LCR meter (E4980A, Agilent Technology). The capacitance and loss tangent were measured at frequencies from 1 kHz to 2 MHz. The measurement results are shown in **Figure 11**. The electrostatic capacity of VEHs were 1.6 nF (sample 1) and 1.8 nF (sample 2) at 1 kHz, respectively. The calculated dielectric constants were approximately 16 and 10, respectively. These dielectric constants are in the same order as the reported values such as 13 at 1 kHz [25]. Loss tangents were 0.013 (sample 1) and 0.021 (sample 2) at 1 kHz, respectively. These values are equivalent to the previous reported PVDF film properties [26].

4.2. Vibration test and power generating result

One end of the plate was clamped by a fixture to form a cantilever. A proof mass (m = 0.2 g) was attached to the other end. Effective length of clamped cantilever was 25 mm. **Figure 12** shows a schematic illustration and photograph of fabricated cantilever VEH.

Figure 13 shows measurement setup of vibration test. Fabricated VEH was mounted on a vibration generator (PET-05, IMV Corp.). Lead lines between top and bottom electrodes were connected to the external load resistance R, which is composed of a fixed resistance (100 kΩ) and a variable resistance (from 0 to 9 MΩ) in series. A laser displacement meter was used

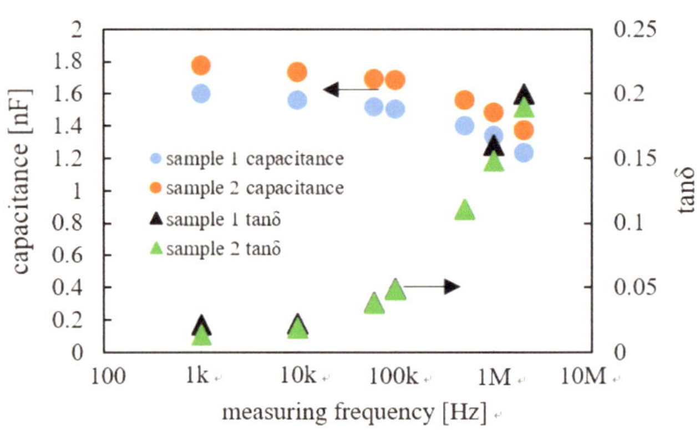

Figure 11. Capacitance and loss tangent of the VEH device.

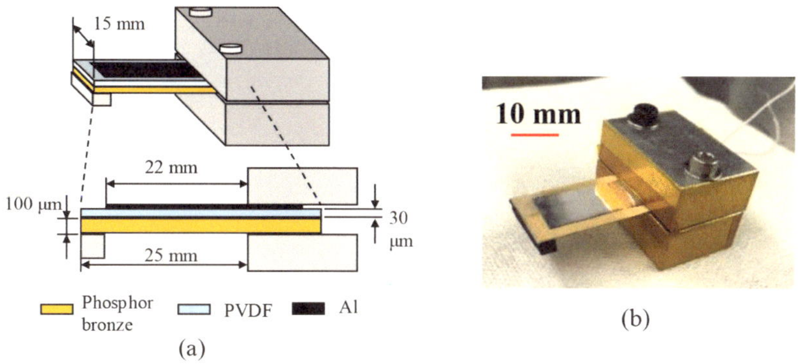

Figure 12. (a) Schematic illustration and (b) photograph of VEH using bar-coated PVDF film on phosphor bronze cantilever.

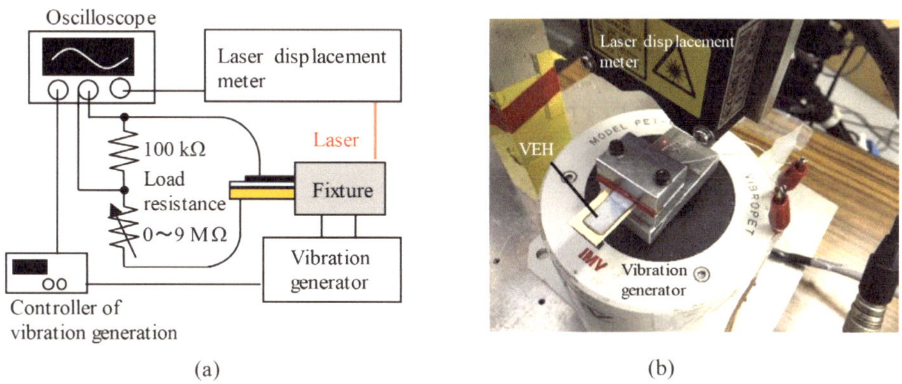

Figure 13. (a) Schematic illustration and (b) photograph of experimental setup for verifying power generation for free vibration.

for determining the acceleration. Output voltage $V(t)$ generated at the load resistance was measured by an oscilloscope. Output power P was calculated by Eq. (1).

$$P = \frac{\int p(t)\,dt}{T}, \tag{1}$$

where $p(t) = V(t)^2/R$, T is the period. By changing R from 0.1 to 9.1 MΩ at intervals of 1 MΩ, the optimal value was searched.

Acceleration and frequency of vibration were set to 17 m/s^2 and 55 Hz (it is the resonant frequency of the cantilever), respectively. At the same time, state of vibrations was taken using a high-speed camera (Memrecam fx-k5, Nac Corp.). In this vibration conditions, half amplitude of the free end of cantilever was approximately 6 mm, as shown in **Figure 14**.

Measurement result of the vibration test is shown in **Figure 15**. Maximum output of 4.3 µW was achieved at R = 2.1 MΩ in case of sample 1. Looking at this result, the thicker PVDF film is, the higher output power is obtained. The reason is considered as follows: in the thicker

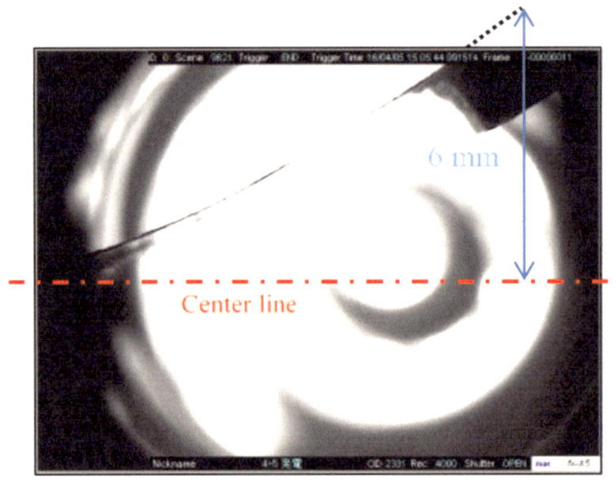

Figure 14. Image of the vibration of the VEH taken by high-speed camera.

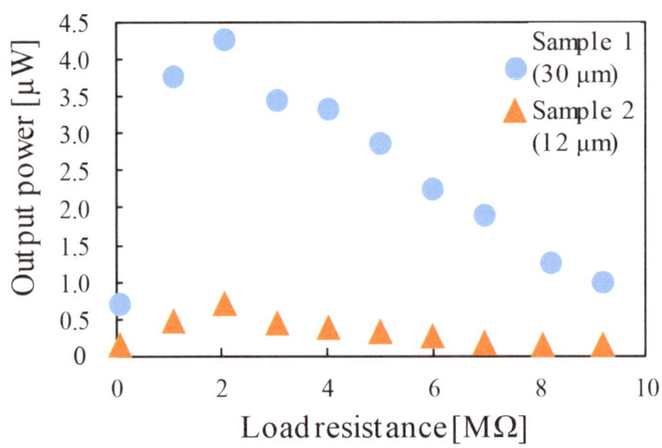

Figure 15. Output power vs. load resistance for free oscillation at 17 m/s^2, 55 Hz.

PVDF film of VEH (sample 1), the distance between neutral axis and PVDF film is greater than that in the thinner PVDF film of VEH (sample 2). Therefore, stress and strain induced on the PVDF film of sample 1 are higher than those of sample 2. In simple and approximately calculation, the 30/12 = 2.5 times larger strain would lead to $2.5^2 = 6.25$ times larger output power. Looking at **Figure 15**, the peak power of sample 1 is 4.3 μW and the peak power of sample 2 is 0.7 μW; the former is 6.1 times larger than the latter, which agrees with the calculation.

Table 3 shows the performance of our VEH and comparison with other cantilever-type piezoelectric VEHs. According to the reports by Kanno et al., the output of their VEHs composed of ceramic PZT film achieved 1–17 μW [6–10]. However, the vibration frequency of power generating test was 108–890 Hz. This value is higher than our VEH. Considering that the theoretical output power is proportional to the square of frequency, our VEH composed of piezoelectric polymer PVDF film shows performance equivalent to that of other VEHs using ceramic PZT.

Although piezoelectric constant of PVDF is lower than PZT [27], we could obtain equivalent power. This is caused by large deformation due to flexible polymer PVDF, which increases power generation [28].

	Piezoelectric material	Effective volume [mm^3]	Frequency [Hz]	Accerakation [m/s^2]	Out put power [μW]	Power density [μW/mm^3]
I. Kanno[8]	PZT	0.2	344	25	2.5	12.5
Sung Sik Won [11]	KNN	2	132	9.8	3.62	1.8
G. Tang [12]	PMN-PT	0.6	237	19.6	5.93	9.83
Z.Cao [28]	AlN	1.7	69	15.7	8.7	5.18
Q.Wang [29]	PZT	4.3	89	9.8	15.4	3.62
This study	PVDF	6.6	55	17	4.3	0.65

Table 3. Comparison of VEHs performance.

5. Spray-coating method

5.1. Spray-coating equipment

To coat PVDF thin film onto the 3D surface, spray-coating method was carried out. **Figure 16** shows the setup of spray-coating equipment (Ushio Mechatronics, Inc.: USC-2114SM-ANG) used for this study. The spray nozzle was set in the chamber facing down toward a work stage. A sample was placed on the stage, which was computationally controlled to move at a constant speed of 60 mm/s. The sample was fixed on the stage by a vacuum chucking. The stage moves under the spray nozzle so as to draw a zigzag course repeatedly (240 times), allowing uniform PVDF coating on the sample. The PVDF solution and N_2 gas (pressure: 0.4 MPa) were supplied to the nozzle through individual tubes, and these fluids were physically mixed in it, followed by sprayed in the form of circular cone having tip angle of 24° (see **Figure 16**). Flow rate was controlled at 60 g/h using a mass flow controller. The distance between the nozzle and the work stage was set to be 45 mm. The total time of spray-coating process was 20 min.

There are several parameters to be set for spray-coating method [29]. These parameters have a complex effect on film properties to each other. Among the parameters, in this study, the kind

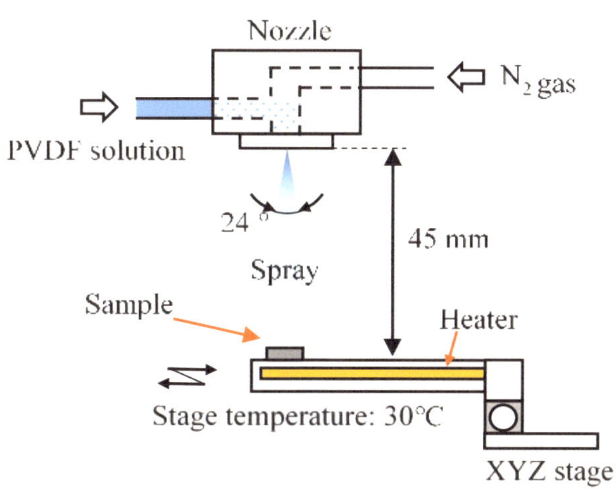

Figure 16. Schematic illustration of the setup of spray-coating equipment.

of organic solvent of diluting PVDF and its dilution rate were changed and examined. The other parameters such as flow rate were set at default values as mentioned above.

5.2. PVDF thin-film formation by spray-coating method

PVDF solution which is already dissolved by vendor in 24 wt% with methyl ethyl ketone (MEK) was used. However, this solution cannot be used for spray coating because of its high viscosity. To adjust the concentration of PVDF, the original PVDF solution was further diluted with organic solvent in addition [30].

At first, PVDF solution was diluted with additional MEK and spray-coated on flat surface of a silicon substrate. The concentration of PVDF after additional dilution was as follows: (a) 1.2 wt%, (b) 1.8 wt%, (c) 2.4 wt%, and (d) 3.0 wt%. After spray coating, PVDF thin film was baked at 90° on a hot plate for 10 min to evaporate the organic solvent (MEK) completely.

The thickness and surface morphology were measured using a contact-type stylus step profiler. To measure the thickness, a part of PVDF thin film on a silicon substrate was cut and a level difference was provided.

Figure 17 shows results of measuring surface morphology with surface roughness (R_a) on each condition. Surface roughness R_a was calculated from the following equation:

$$R_a = \frac{1}{L}\int_0^L |y(x)|dx \tag{2}$$

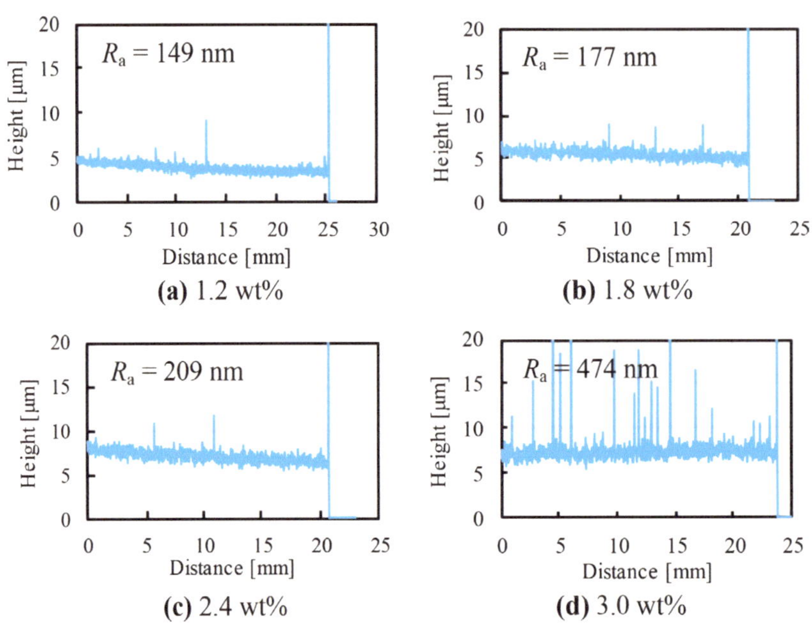

Figure 17. Surface morphology and roughness of spray-coated PVDF thin films measured by a stylus step profiler. Weight ratio of PVDF to total weight varies for each sample.

where L is the evaluation length and $y(x)$ is the height along x direction. These measurement results showed that the higher concentration of PVDF, the thicker PVDF film was obtained. However, as the concentration of PVDF increases, surface roughness R_a increases. In case of condition (d) 3.0 wt%, many particles of which size is approximately 10 μm were observed. Several particles were also observed under other conditions. **Figure 18** shows a SEM image in case of condition (c) 2.4 wt%, in which many particles are observed. They may be PVDF particles generated by the spray-coating process. The drying speed of solvent of diluting PVDF may affect generating particles. In case of only using MEK as diluting solvent, PVDF is sprayed to the sample in a solid state. It is because the boiling point of MEK is low, which promotes its evaporation before arriving at the sample. These particles of generating several micron surface roughness may produce bad effects on the electrode formed on PVDF film. Since the electrode thickness is usually less than 1 μm, the particles may cause the crack of electrode, which leads to the reduction of effective electrode area. Improving the surface condition of spray-coated film is future study.

5.3. Spray-coating PVDF on 3D surface

PVDF solution was spray-coated on a helical compression spring. A helical spring was adopted because it can be largely deformed. Since PVDF can easily follow the deformation of helical spring due to its flexibility, high output power is expected. The material of spring was stainless (SUS304), and its size is as follows: free length: 4.3 mm; coil diameter: 2.16 mm; and wire diameter: 0.16 mm.

Figure 19(a) shows the schematic of the setup of spray coating for 3D helical spring. **Figure 19(b)** shows a SEM image of spray-coated PVDF thin film coated on it. The SEM image shows that the PVDF film was formed even on the bottom side of helical spring, showing that the PVDF solution was spray-coated omnidirectionally in this case. Eventually, PVDF thin

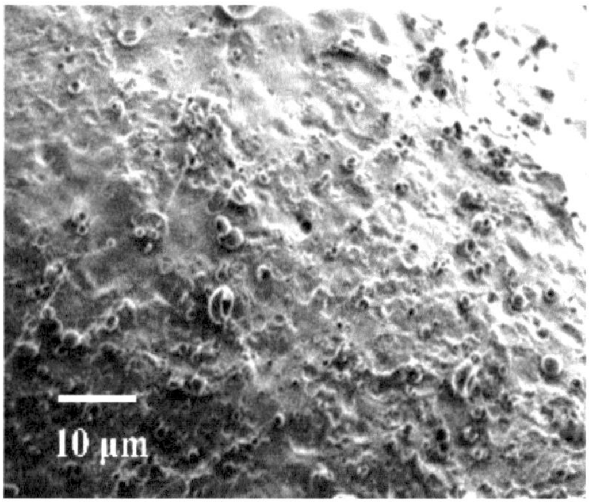

Figure 18. A SEM image of the surface of spray-coated PVDF thin film formed by the condition (c).

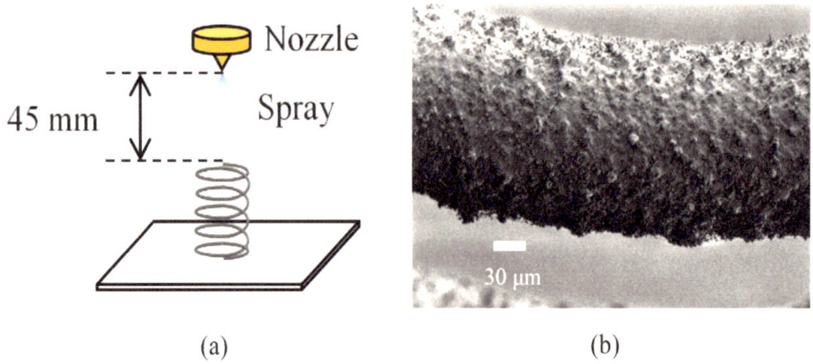

Figure 19. (a) Setup of spray coating for helical spring and (b) a SEM image of the surface of PVDF thin film.

film deposition to the helical spring was successfully achieved. Energy harvesters composed of PVDF-coated 3D spring such as helical spring will be fabricated in the future.

6. Conclusion

PVDF film was formed by bar-coating method and characterized its mechanical and piezoelectric properties. A cantilever-type vibration energy harvester (VEH) made of a PVDF film was fabricated on a phosphor bronze rectangular plate. The dimensions of the cantilever were 25 mm in length and 15 mm in width. After corona polarization, the piezoelectric constant d_{33} of PVDF was calculated to be −8.5 pC/N and −10.1 pC/N at 30 and 12 μm thickness, respectively. As a result of the vibration test, the maximum output at $R = 2.1$ MΩ was 4.3 μW in case of the device with a thickness of 30 μm, which was the same performance as other cantilever-type VEH using ceramic piezoelectric PZT. PVDF is a flexible material and it was able to generate large strain due to flexibility of the device.

The thickness of PVDF film and corona polarization condition should be optimized for improving piezoelectric constant of PVDF film. Optimizing the structure of VEH should be done using MEMS technology in the future.

PVDF thin film was formed by spray-coating method. PVDF thin film was deposited on a 3D surface by spray coating. A helical compressing spring was adopted. The deposition on it was successfully achieved. However, surface conditions of it are currently not good. Improving the surface conditions of PVDF film and its application to a power-generating device are ongoing works.

Acknowledgements

This study was partially conducted by MEXT*-Supported Program for the Strategic Research Foundation at Private Universities, "Creation of 3D nano-micro structures and its application to biomimetics and medicine," 2015–2019 (*MEXT: Ministry of Education, Culture, Sports, Science and Technology, Japan).

We would greatly thank Prof. Kozuka at Kansai University for his advising and helping in measurement of XRD pattern and residual stress of PVDF thin film.

We would greatly thank Daikin Industries, Ltd. for providing the PVDF solutions.

Author details

Hiroki Takise*, Masato Suzuki, Tomokazu Takahashi and Seiji Aoyagi

*Address all correspondence to: aoyagi@kansai-u.ac.jp

Kansai University, Suita, Osaka, Japan

References

[1] Beeby SP, Tudor MJ, White NM. Energy harvesting vibration sources for microsystems applications. Measurement Science and Technology. 2006;**17**(12):175-195

[2] Matsumoto K, Saruwatari K, Suzuki Y. Vibration-powered battery-less sensor node using MEMS electret generator. Digest Tech. Power MEMS'11. 2011:134-137

[3] Hosokawa S, Hara M, Oguchi H, Kuwano H. Vibration based energy harvester employing ZnO film on the stainless steel substrate. IEEJ Transactions on Sensors and Micromachines. 2013;**133**(4):126-127

[4] Liu JQ et al. A MEMS-based piezoelectric power generator array for vibration energy harvesting. Microelectronics Journal. 2008;**39**(5):802-806

[5] Roundy S, Wright PK. A piezoelectric vibration based generator for wireless electronics. Smart Materials and Structures. 2004;**13**(5):1131-1142

[6] Tsujiura Y, Adachi K, Kanno I. Piezoelectric MEMS energy harvesters of PZT thin films on stainless steel cantilevers. PowerMEMS'12. 2012:500-503

[7] Kanno I, Morimoto K, Wasa K, Kotera H. Energy harvester of c-axis oriented epitaxial PZT films on stainless steel sheets. IEEJ Sensors and Micromachines. 2009:594-597

[8] Kanno I, Sagawa K, Oka R, Kotera H, Ogawa J, Yamauchi N, Aizawa K, Matsushima T. Piezoelectric energy harvesters of PZT films deposited on titanium cantilevers. Power MEMS'10. 2010:379-382

[9] Kanno I, Kotera H, Shibata K, Mishima T. Characterization of vibration energy harvesters composed of piezoelectric thin films. The Japan Society of Mechanical Engineers Micro-Nano Science & Technology Division. 2011:26-27

[10] Tsujiura Y, Suwa E, Kurokawa F, Hida H, Kanno I. Reliability of vibration energy harvesters of metal-based PZT thin films. Power MEMS'14. 2014:557

[11] Won SS et al. Lead-free Mn-doped ($K_{0.5}$,$Na_{0.5}$)NbO_3 piezoelectric thin films for MEMS-based vibration energy harvester applications. Applied Physics Letters. 2016;**108**:232908

[12] Tang G et al. Piezoelectric MEMS low-level vibration energy harvester with PMN-PT single crystal cantilever. IET Electronics Letters. 2012;**48**(13):784-786

[13] Toprak A, Tigli O. MEMS scale PVDF-TrFE-based piezoelectric energy harvesters. Journal of Microelectromechanical Systems. 2015;**24**(6):1989-1997

[14] Sasaki M et al. Heating effect on photoresist in spray coating technique for three-dimensional lithography. Japanese Journal of Applied Physics. 2003;**42**(Part 1, No. 6B): 4027-4030

[15] Fujitsuka N et al. Monolithic pyroelectric infrared image sensor using PVDF thin film. Sensors and Actuators A. 1998;**66**:237-243

[16] Oh SR et al. Residual stress in piezoelectric poly(vinylidene-fluoride-co-trifluoroethylene) thin films deposited on silicon substrates. Thin Solid Films. 2010;**519**(4):1441-1444

[17] Choi M et al. Mechanical and electrical characterization of PVDF-ZnO hybrid structure for application to nanogenerator. Nano Energy. 2017;**33**:462-468

[18] Porter DA et al. Effects of in-situ poling and process parameters on fused filament fabrication printed PVDF sheet mechanical and electrical properties. Additive Manufacturing. 2017;**13**:81-92

[19] Liu W, et al. Fabrication of Spiral-Shaped PVDF Cantilever Based Vibration Energy Harvester. In: Proc. IEEE, Nano/Micro Engineered and Molecular Systems; 2014. pp. 458-461

[20] Choi SB et al. Measurement of flexoelectric response in polyvinylidene fluoride films for piezoelectric vibration energy harvesters. Journal of Physics D: Applied Physics. 2017;**50**:075502

[21] Hakko co., Ltd. Available from: http://www.hakko.co.jp/qa/qakit/html/h01100.htm

[22] Furukawa T, Goho T, Date M, Takamatsu T, Fukada E. Piezoelectricity of corona-poled poly (vinylidene fluoride). Kobunshi Ronbunshu. 1979;**36**(10):685-688

[23] Measurement Specialties Inc. Piezo Film Sensors, Technical Manual. Online Available. Available from: http://www.meas-spec.com

[24] Dargahi J. A piezoelectric tactile sensor with three sensing elements for robotic, endoscopic and prosthetic applications. Sensors and Actuators. 2000;**80**:23-30

[25] Sasaki Y et al. Correlation between different crystal structures and physical properties of PVDF films fabricated by solvent casting from a single solvent. Kobunshi Ronbunshu. 2013;**70**:489-495

[26] Jain A et al. Dielectric behaviour of PVDF thin films. Indian Journal of Advances in Chemical Science. 2014;**2**(3):212-216

[27] Tanaka T, Murakami S, Uno M, Yamashita K. Development of MEMS ultrasonic sensor using P(VDF/TrFE) thin films. IEEJ Transactions on Sensors and Micromachines. 2015;**135**(5):145-151

[28] Cao Z, Zhang J, Kuwano H. Design and characterization of miniature piezoelectric generators with low resonant frequency. Sensors and Actuators A: Physical. 2012;**179**:178-184

[29] Wang Q, Cao ZP, Kuwano H. Metal-based piezoelectric energy harvesters by direct deposition of PZT thick films on stainless steel. Micro & Nano Letters. 2012;**7**(12):1158-1161

[30] Sasaki M et al. Spray coating of photoresist for three dimentional micromachining. IEEJ Transactions on Sensor and Micromachines. 2002;**122**(5):235-243

Chapter 6

Piezoelectric Sensors Used for Daily Life Monitoring

Hirokazu Madokoro

Additional information is available at the end of the chapter

http://dx.doi.org/10.5772/intechopen.77724

Abstract

This chapter presents an unrestrained and predictive sensor system to analyze human behavior patterns, especially in a case that occurs when a patient leaves a bed. Our developed prototype system comprises three sensors: a pad sensor, a pillow sensor, and a bolt sensor. A triaxial accelerometer is used for the pillow sensor, and piezoelectric elements are used for the pad sensors and the bolt sensor that were installed under a bed mat and a bed handrail, respectively. The noteworthy features of these sensors are their easy installation, low cost, high reliability, and toughness. We developed a machine-learning-based method to recognize bed-leaving behavior patterns obtained from sensor signals. Our prototype system was evaluated by the examination with 10 subjects in an environment representing a clinical site. The experimentally obtained result revealed that the mean recognition accuracy for seven behavior patterns was 75.5%. Particularly, the recognition accuracies for longitudinal sitting, terminal sitting, and left the bed were 83.3, 98.3, and 95.0%, respectively. However, falsely recognized patterns remained inside of respective behavior categories of sleeping and sitting. Our prototype system is applicable and used for an actual environment as a novel sensor system without restraint for patients.

Keywords: machine learning, invisible sensors, bed monitoring, piezoelectric elements, and signal processing

1. Introduction

Along with the longevity in our society, labor shortages will be severe, especially at hospitals, nursing homes, and nursing care facilities [1]. Currently, few caretakers care for numerous elderly patients. For this situation, nurses and caretakers should monitor them inadequately, especially during sleep at night [2]. As related to this situation, Mitadera and Akazawa addressed that approximately half of falling or tumbling accidents are among elderly patients [3]. Numerous accidents occurred when elderly patients leave their bed and these accidents occurred in a place where nurses and caretakers are hard to keep their eye on. Statistically, the tumbling and falling accident rates were 83.3% and 85.4%, respectively, when they had no support at an accidental moment. For protecting it, suitable provision is necessary after conducting assessments for respective elderly patients. Actually, one approach is to use bed-leaving sensors that signal when they leave from their beds. The number of hospitals and nursing care facilities using these sensors, which can be used to prevent falling from their beds, has increased recently [4, 5].

As a simple sensor system attached to a patient's clothing [6], clip sensors are the lowest-cost sensors that can be introduced easily. However, the usage of clip sensors has been prevented because of the necessity of wearer constraint, especially for the protection of human rights. Moreover, malfunctions and anomaly detections occur frequently for clip sensors because of the binary output used to detect bed-leaving behaviors. Regarding the reliability and perspective of management, clip sensors are insufficient to prevent falling from a bed completely. Furthermore, accidents caused by cable binding to the neck have been reported as clinical serious accidents [6]. Therefore, clip sensors are regarded as inadequate for using clinical sites, although it is easy to introduce them as a cost-effective solution.

Mat sensors are also widely used as a cost-effective and convenient sensor that can be installed easily [7]. Using mat sensors, Haruyama et al. developed an alarm system to detect patients leaving from their beds [8]. Medical and welfare suppliers have already released numerous models and various types of consumer used mat sensors installed on a floor, a bed, or on rolling handrails. For authentication, mat sensors used on a floor are unnecessary used for medical devices under the pharmaceutical law. Although mat sensors are easy to produce and to sell, a delayed problem remains as detection accuracy because of the response after sitting at the end of the bed. Moreover, sensor responses are apparent when medical staff members such as nurses or physicians walk on. Therefore, it is a challenging task to distinguish the responses of patients and medical personnel for signal pattern recognition.

The sensors that are rolled over handrails not only obstruct a view of a bed, but also present a removal risk when a patient finds it and feels negatively about being restrained. Moreover, false detection occurs when patients leave their bed without gripping a sensor installed handrail. Mat sensors installed on a bed under a sheet can detect bed-leaving behavior patterns with superior reliability to mat sensors of other types. Nevertheless, existing mat sensors are actuated by a binary response similar to that for clip sensors. Early detection is not feasible, especially in the initial bed-leaving behavior.

Using a camera as a bed monitoring sensor system provides a cost-effective solution combined with state-of-the-art computer vision technologies and algorithms that can use as open source software. Although camera can obtain much information for a monitoring subject, it is a challenging task to predict behavior patterns obtained from images as pixel signal information. It is impossible to monitor numerous subjects simultaneously and is unrealistic for a few operators or medical staff members if they observe images directly. Moreover, they must consider the aspect of human rights and quality of life (QoL). Because behavior patterns differ among people [9], it is impossible to recognize bed-leaving behavior patterns using only sensor responses with detailed analyses. Moreover, vision-based monitoring using a camera imposes a mental load on patients because they feel as though they are under surveillance all daytime and nighttime.

For solving abovementioned problems of various types, this chapter presents an unrestrained and cost-effective sensor system regarding QoL. The proposed system can analyze predictive behavior patterns that occur when a patient leaves from their bed. A machine-learning-based method was developed to obviate the setting of thresholds in advance. Moreover, an integrated system was developed for sending output signals obtained from sensors to a monitoring computer using a close-range wireless module. For this chapter, a novel multiple-sensor system is proposed to improve accuracy according to a usage of a subject. The amount of data was reduced to use predicting and minimizing incorrect recognition which is given to a minimum number of sensors. Our sensor system was evaluated in an environment that represents a clinical site. Experimentally obtained results demonstrated that our method predicted bed-leaving behavior patterns of seven types, especially for longitudinal sitting and terminal sitting.

2. Related studies

Hatsukari et al. developed a sensor system to detect bed-leaving three major patterns using strain gauges installed inside of actuators to obtain weight changes of a person on a bed [15]. For their system, three actuators are used in a bed. Each actuator has four biaxial strain gauges installed on diagonal lines. This is the most popular bed-leaving sensor system in practical use currently. As a new product of their company Paramount Bed Co. Ltd., they embedded sensors and a controller to their beds of the highest models. The detection accuracies for longitudinal sitting and terminal sitting were 87.0 and 98.1%, respectively [15]. This performance measure addressed that this system has both high performance and reliability. However, an important limitation of this system is that it is used only for the lifting beds of that company's products with installed actuators. Moreover, it is necessary to install sensors into actuators at the bed manufacturing process. Therefore, users must buy this product if they use this bed-leaving detecting function. Moreover, this sensor system is not casually used, especially for a use case that can be installed later. In addition, strain gauges have an uncertain response to thermal changes. For a characteristic gap caused by thermal changes, they described that they accommodated it with software. Furthermore, this system requires the initial weight setting of a subject to be selected from three divisions in advance.

Using a watch-type triaxial acceleration sensor, Iomoto et al. [12] measured body motion patterns during sleeping as quality of sleep (QOS). They detected body motion patterns at 1 min intervals in band-pass signals of 2–3 Hz acceleration changes using the threshold of ± 0.01 G. They recognized sleeping or arousal for time-series active mass datasets. Their proposed method was evaluated for 1 week using datasets obtained from six subjects in their 50s who were diagnosed as needing daily life guidance. The experimentally obtained results revealed that each subject assessed sleeping parameters quantitatively. However, the results contained a shortcoming to demonstrate a global tendency in light of relations to temporary arousals during sleep. Moreover, this watch-type sensor decreases QOS when worn on an arm during sleeping because of the necessity for a subject to be restrained.

As an unrestrained monitoring sensor system, Okada et al. proposed a method to install an acceleration sensor in a comforter cover [10]. This system was constructed to measure the heart rate of a subject during sleeping. Near the position of the heart in a subject, this sensor was stitched to the surface of a cover on the upper-left chest region. The piezoelectric ceramics-based acceleration sensors have advanced sensitivity. They obtained measurement results resembling electrocardiogram (ECG) waves during face-up sleeping under resting status. The acceleration signal peak was obtained in the corresponding position on the ECG. Moreover, the signal peaks, which were equivalent to peaks of the R wave, were counted to assess the side sleeping behavior. The system with a compact sensor of 4 g weight requires amplifier circuits and convertors from analog signals to digital signals with capturing and processing in a computer. However, numerous details are remained in their system in terms of sensor wiring with a comforter cover as a remaining obstacle to practical use of this device.

Bed-leaving sensors and other systems of several types were proposed by Madokoro et al. [19–23] as simple and unrestrained sensors that can be installed in existing beds. Their first prototype was pad sensors [19] installed between a bed and a cover to detect behaviors of a subject globally. For this system, piezoelectric films were used for the bound with acrylic resin boards to detect pressure for measurement of the upper body movements of a subject on a bed. Their second prototype was pillow sensors [20] for measurement of the upper body movements of a subject on a bed. For this system, a three-axis gyroscope protected with a small capsule was used for installation into a pillow. Their third prototype was bolt vibration sensors [18] installed to a safety railing beside a bed. The size of bolt vibration sensors [17] was downsized because of the use for measuring bridge vibration. Using these three prototypes, we proposed multiple sensor systems [21] concomitantly with use cases and the balance of cost and accuracy. Our latest prototype was caster sensors [22] used for measuring the weight distribution of a bed.

However, several problems persisted in terms of a slip or drift of the pad sensors and the pillow sensors. No response is apparent for bolt sensors without gripping of the safety railing by a patient. Moreover, a medical doctor at a hospital made the comment that caster sensors present difficulties for moving a bed in a ward. Therefore, it is necessary to develop a new sensor that can resolve these problems, not merely with a slip and drift, but also for a frequently moving bed.

3. Sensors and their integration system

Various functional and high-performance sensors were proposed for existing bedside monitoring systems with targeting expensive care or medical treatments [3]. Alternatively, attaching them to the body is necessary to realize steady and long-term monitoring [12, 10]. For this chapter, a sensor system is designed while being practical, convenient, inexpensive, and simple.

Particularly, the following features are examined as (1) a bed mat to distribute the body weight patterns, (2) a pillow to measure head weight patterns, and (3) a handrail which is gripped by a subject while standing immediately before leaving. For multiple sensing, our system comprises following three sensors: (1) a pad sensor under the bed mat, (2) a pillow sensor to detect acceleration according to head movements, and (3) a bolt sensor used for a handrail joint. The features in each sensor and the overall system architecture are described as the following.

3.1. Pad sensor

Our originally developed pad sensors are easy to install under a bed sheet. **Figure 1(b)** depicts the design structure and an overview picture of our prototype sensors. The shape of pad sensors is round because they can detect weight changes from all directions. Piezoelectric films by Measurement Specialties Inc. [11] were used for this sensor prototype. The piezoelectric films are fixed between two polyethylene terephthalate (PET) plates of laminated polyester. The polyester and PET plate sizes were, 125 μm and ϕ 70 $\times$ 0.5 mm, respectively.

Output signals are generated from the bent piezoelectric films when a subject transfers body weight on the bed. This sensor is possible to measure recursively because the reference potential is offset when the bending stops. Moreover, the weight strength according to body changes is obtainable linearly because the bending angle of the piezoelectric film and output voltage has a relation of proportionality. Herein, piezoelectric films are less troublesome and provide no false operations because of simple wiring without electric power supply cables for measurements. Therefore, a cost-effective system can be provided as requiring no maintenance related to replacement of a battery.

3.2. Pillow sensor

For the pillow sensor, a triaxial acceleration sensor module MMA7361LC produced by Freescale Semiconductor Inc. is used. **Figure 1(b)** presents a photograph of the device installed in the case. As shown on the left side of the photograph, after storing the device, a cap for sealing was inserted. Herein, basic characteristics of this sensor module were evaluated by our former study of attitude control for a flying robot [16].

A lithium-ion polymer battery with 0.85 Ah capacity is used as the power source of this sensor. Additionally, an original case was designed for storing the board and the battery to a pillow. Using a three-dimensional printer, the cylindrical case was made from acrylonitrile butadiene styrene (ABS) with the size of 55 mm diameter and 105 mm long. The board and battery were

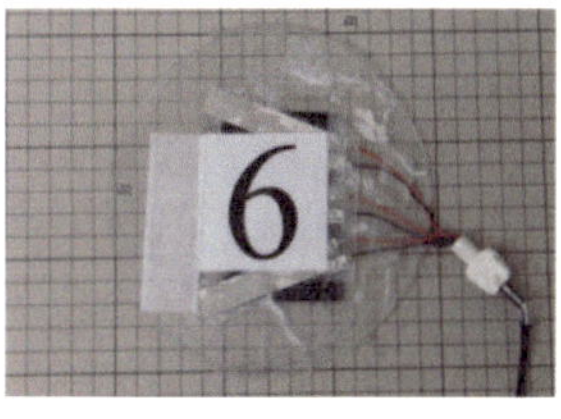

(a)

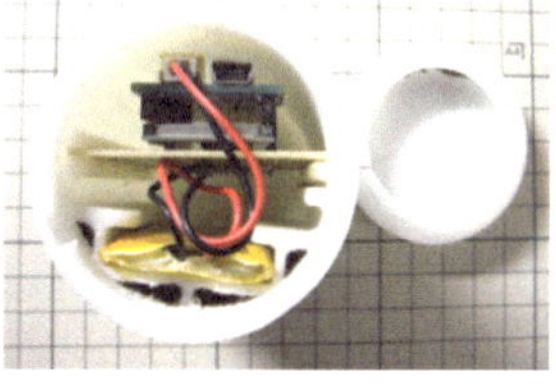

(b)

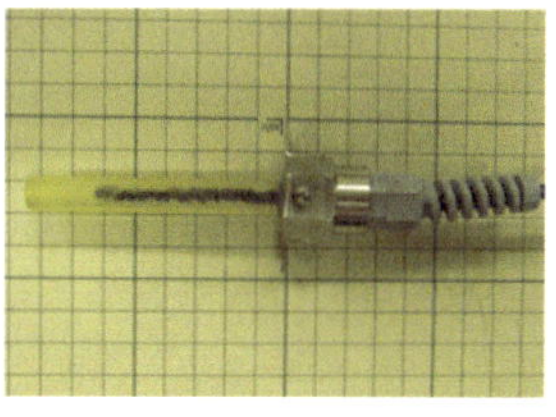

(c)

Figure 1. Developed sensors: (a) pad sensor, (b) pillow sensor, and (c) bolt sensor.

stored to the upper and lower part in the case using sliding rails without a margin that is set for the slides because the sensor boards were fixed inside the case.

3.3. Bolt sensor

A bolt sensor is used for the joint between a bed and a handrail. This sensor performs overall bed sensing, whereas pad sensors are used for partial subject sensing. Weight changes are detected with this sensor that occurs when a subject tries to stand up and grip a handrail. **Figure 1(c)** depicts our bolt sensor prototype that was developed originally by Shimoi et al. [17] as a simple and smart sensor to measure bridge strain. His bolt sensor has been commercially available for some time. Based on his bolt sensor, a novel sensor was originally developed with ϕ =10 mm to fit the size of the bolt to the joint of the bed used for our evaluation experiment.

A piezoelectric cable was inserted into hollow polyurethane resin. Subsequently, a bolt-shaped adapter was attached as a cover for the exterior. Using this adapter, the bolt sensor can be fixed to a nut. Although no electric power source is necessary for piezoelectric cables, electric wires were extracted to obtain sensor signals. The data sheet denotes that piezoelectric cables generate voltage linearly according to bending angles. This linear feature was reported by Shimoi et al. [17] through a preliminary experiment and earlier experiments.

3.4. System structure

Figure 2(a) portrays the entire structure of our sensor system that comprises six pad sensors installed over the bed, one bolt sensor mounted on the joint part between a handrail and a frame of the bed, and one pillow sensor using a triaxial acceleration module installed inside of a pillow.

The sensor installation positions are annotated from S1 to S6. The assignment of six pad sensors was S1 and S2 for the shoulder part, S3 and S4 for hip part, and S5 and S6 for the terminal part. These six sensors were installed with referring to the literature related to the monitoring system development used for welfare care beds [4, 5]. A bolt sensor was mounted

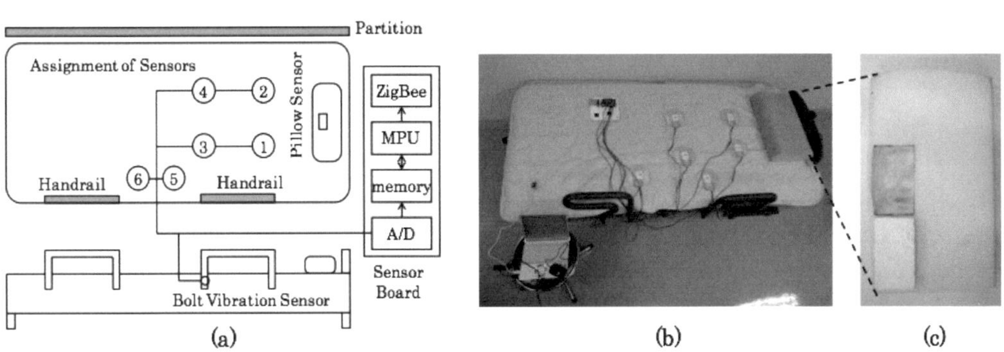

Figure 2. Block diagram of our proposed system and assignment of our developed sensors on the bed. The symbols of 1–6 correspond to the sensor assignment of S1–S6.

to the joint of a handrail on one side of the bed. The size of the bed used for this experiment is 2100 mm long, 1000 mm width, and 700 mm height. Although the bed equips an actuator used for back-panel reclining, this function was not used for the experiment.

A commercially available pillow was used for embedding the sensor. This low-resilience urethane pillow purchased from Nitori Co. Ltd. was 500 mm wide, 310 mm deep, with 90 mm height. The pillow backside was cut out as shown in **Figure 2(c)** to produce a void space to insert the sensor case. The sensor case was fixed with three points using a hook and loop fastener to prevent rolling inside the pillow. One advantage of our sensor system is its easy installation as an aftermarket add-on part. Herein, it is necessary to open a pillow and partially empty it during repairs.

The sensor system requires no calibration. Regarding the place to store the sensor, it would feel uncomfortable or decrease the cushioning feature when putting the head to the pillow. Participated subjects in the experiment reported that they felt no negative impression from using this pillow.

The primary purpose of our system is to realize real-time monitoring using unrestrained and smart sensors with keeping QoL, which is important for a patient to live life normally. Our proposed system requires no supervision using infrared cameras or constraining sensors such as collar clip sensors. Moreover, a cost-effective system is concerned using piezoelectric films and cables as sensors that can function with minimum trouble or missed operations, and with remarkable duration and characteristics for pressure resistance.

Output voltage signals from each sensor were obtained using a board. The board was developed using Open Source Hardware Arduino FIO with a wireless communication module. With consideration of power consumption, a short-range wireless module of ZigBee was used for the communication with a monitoring terminal computer. Measured signals are displayed to a monitoring computer in real time. The method based of machine learning was embedded as real-time execution software for bed-leaving prediction.

4. Prediction method

4.1. Target behavior patterns

The target behavior patterns for bed-leaving prediction comprise three groups: sleeping, sitting, and leaving. Detailed behavior patterns were classified from output signals of six sensors. The sleeping includes three patterns: face-up sleeping, left sleeping, and right sleeping. The sitting includes three patterns: longitudinal sitting, lateral sitting, and terminal sitting [14]. The total prediction target is to produce seven patterns including leaving. The following descriptions are global features and estimated sensor responses in respective patterns.

Face-up sleeping: sleeping on the bed normally to the upper side of the body.

Right sleeping: rolling over on the bed to the right side.

Left sleeping: rolling over on the bed to the left side.

These three patterns occur as sleeping behaviors. The following patterns are the target of bed-leaving prediction.

Longitudinal sitting: sitting longitudinally on the bed after rising.

Lateral sitting: sitting laterally on the bed after turning the body from longitudinal sitting.

Terminal sitting: sitting in the terminal position on the bed trying to leave the bed. Rapid and correct detection are necessary because of the terminal situation for leaving the bed.

Left the bed: left from the bed.

Therein, sensor output signals disappear in the status of losing consciousness or a life crisis in terms of cardiopulmonary arrest. Monitoring devices in terms of electrocardiographs are used for these circumstances. Therefore, such circumstances are beyond our detection and prediction targets.

Longitudinal sitting is a behavior pattern by which a subject sits up on the bed. In numerous cases, subjects will return to sleeping again. In lateral sitting, a subject will move to leave from their bed because of turning their body to the terminal. For this position, our system should determine lateral sitting immediately to predict a behavior pattern leaving from the bed. Moreover, it is necessary to detect rapidly and correctly because of the terminal situation for leaving the bed if a subject moves to longitudinal sitting. Patients are protected from injury or accidents caused by falling from the bed if our sensor system can detect such phenomena before leaving completely.

4.2. Preprocessing

Frequently, original sensor signals include much noise, which decreases the recognition accuracy and increases calculation costs. As a noise removal method, threshold-based filters have been used widely for preprocessing. In contrast, calibration in advance is necessary according to a monitoring target because thresholds are set empirically or subjectively. Moreover, it is a challenging task to absorb characteristic variation of sensors used only for fixed thresholds. For targeting humans as a measurement objective, individual variations have strong effects.

Various methods have been proposed for machine-learning algorithms. For this chapter, self-organizing maps (SOMs) [27] are used for similarities of input data in the training process. Based on competitive and neighborhood learning, SOMs create clusters using unsupervised learning-based self-mapping characteristics. For practical applications, SOMs are validated as effective and superior performance through existing studies [33]. In the advanced computing performance, SOMs have facilitated various applications such as facial image processing, medical image processing, character recognition, text mining, and remote sensing data analysis.

As a classical approach, *k*-means [24] is widely used for unsupervised learning-based clustering. Vesanto et al. evaluated that SOMs have superior performance to that of *k*-means in their numerical experiments [31]. Moreover, Terashima et al. [29] demonstrated that false recognition accuracy of SOMs decreased to the minimum compared with that of *k*-means used for clustering. Therefore, SOMs are used in our method because of these benefits.

The SOM network comprises two layers: an input layer and a mapping layer. Input layer units are assigned as the number of features of input data. Two units were set for a triaxial acceleration sensor. The mapping layer consists of units in a low-dimensional space. For our method, mapping units were assigned as one dimension because of using it for vector quantization on clustering. The learning process is executed for bursting a unit on the mapping layer when a set of input signals is given.

SOM training algorithms are the following. $w_{i,p}(t)$ are weights from an input layer unit i $(i = [1, I] \in \mathbb{Z})$ to a mapping layer unit j $(j = [1, J] \in \mathbb{Z})$ at time t. These weights are initialized randomly before training. The Euclidean $L2$ distance d_j between $x_i(t)$ and $w_{i,j}(t)$ is calculated as

$$d_j = \sqrt{\sum_{i=1}^{I} \left(x_i(t) - w_{i,j}(t)\right)^2}. \tag{1}$$

The unit for which d_j is the smallest is sought as the winner unit c.

$$c = \operatorname{argmin}\left(d_j\right). \tag{2}$$

The neighborhood region $N_j(t)$ around c is defined as

$$N_j(t) = S\left(1 - \frac{t}{O}\right). \tag{3}$$

Therein, S $(0 < S \leq J)$ is the initial size of $N_j(t)$; O is the maximum iteration for training. Subsequently, $w_{i,j}(t)$ of $N_j(t)$ are updated to close input feature patterns.

$$w_{i,j}(t+1) = w_{i,j}(t) + \alpha(t)\left(x_i(t) - w_{i,j}(t)\right). \tag{4}$$

Therein, $\alpha(t)$ is a training coefficient that has decreasing value with the progress of training as

$$\alpha(t) = \alpha(0)\left(1 - \frac{t}{O}\right), \tag{5}$$

where $\alpha(0)$ $(0 < \alpha(0) \leq 1.0)$ is the initial setting value. In the first stage, the training speed is rapid when the rate is high. In the final stage, the training converges while decreasing the rate.

4.3. Prediction and recognition

After noise removing from originally obtained sensor signals, behavior patterns are recognized using supervised learning-based methods. Supervised learning intends to acquire information representation [30]. In contrast, unsupervised learning intends to create mapping relations among data [26]. Herein, support vector machines (SVMs) are popularly used methods for supervised learning. SVMs provide advanced performance with mapping to a high dimensional space using a kernel function. Moreover, Boosting is a method combining numerous weak learning machines. As a simple algorithm approach, counter propagation networks (CPNs) [25] are used as a supervised learning algorithm to be expanded based on unsupervised learning on SOMs.

CPNs were used as our method, not SVMs or Boosting, because of the following two features: (1) the learning algorithm is easy to implement for inserting a Grossberg layer to SOMs that are used for preprocessing; (2) relations among signals are visualized through the creation process of mapping structures. Formulas are described as learning algorithms on CPNs in different parts of SOMs. Visualization of relations among signals is actualized on a category map [33]. Particularly, it presents visualization of mapping results obtained as figures in the evaluation experiment.

CPNs, which were proposed by Nielsen [25], contain the network structure to append a Grossberg layer as the third layer to be supplied teaching signals. The input and mapping layers of CPNs are similar to those of SOMs. The Grossberg layer is assigned to the counter position of the input layer. For this chapter, a two-dimensional mapping layer was selected for visualization of similarity among features of input data.

The CPN training algorithm from presenting input data through updating weights after searching the winner unit c consists of the similar procedure of SOMs. However, weights between the input layer and the mapping layer and neighborhood regions are changed respectively to $w_{i,j,k}(t)$ and $N_{j,k}(t)$ because of the use of a two-dimensional mapping layer. The $v_{j,k,l}(t)$ are weights from a Grossberg layer unit l ($l = [1, L] \in \mathbb{L}$)) to a mapping layer unit (j, k) at time t. $v_{i,j,k}(t)$ and its neighborhood units inside $N_{j,k}(t)$ are updated based on the following Grossberg-learning algorithm.

$$v_{j,k,l}(t+1) = v_{j,k,l}(t) + \beta(t)\left(T_j(t) - v_{j,k,l}(t)\right), \tag{6}$$

where T_j represents teaching signals. Similarly to $\alpha(t)$, $\beta(t)$ is a training coefficient that decreases its value with the progress of training as

$$\beta(t) = \beta(0)\left(1 - \frac{t}{O}\right), \tag{7}$$

where $\beta(0)$ $\left(0 < \beta(0) \leq 1.0\right)$ is the initial value. Finally, as the maximum value of $v_{j,k,l}(t)$ for the Grossberg layer unit l, label $S_l(t)$ is searched for the following.

$$S_j(l) = \underset{1 \leq l \leq L}{\text{argmax}} \left(v_{j,k,l}(t)\right). \tag{8}$$

After labeling all units on a mapping layer, a category map is automatically created as a training result. Subsequently, test datasets are given to CPNs. A mapping layer unit is bursted as the minimum value of the Euclidean $L2$ distance in the formula (1). Recognition results are presented as corresponding labels.

5. Sensor characteristics evaluation

5.1. Setup

For evaluating characteristics of our developed film-load sensors, preliminary experiments were conducted using the load test machine (Multi Force Analyzer FWT-1000; DigiTech Co. Ltd.). The major specifications of the machine are as follows: 1 kN rated weight; 100 mN resolution; 600 mm/min maximum test speed; and ±0.2% weight precision. The majority of loads are attained from the vertical side as a surface load because of the installation of the sensors on the bed frames. For this load test, a fixture was made of A2017 duralumin, as depicted in **Figure 3(a)**. The major specifications of the fixture are 100 × 100 mm with 15 mm basement thickness and 70 × 50 mm with 5 mm top thickness.

The output characteristics were evaluated using our developed sensors of five sets with the default test speed of 5 mm/min. **Figure 3(b)** depicts a schematic diagram of the sensor output that occurs from the range except that of the rivet parts. For attaining a load, the sensor is fixed to the removal part of 10 mm from the boundaries. The output voltages are measured from respective sensors using a data logger (LR8431; Hioki Corp.) concomitantly with the test load.

5.2. Basic sensor characteristics

Figure 4 depicts the output characteristics of five sensors. The vertical and horizontal axes, respectively, depict output voltage and test loads. The output voltage increases concomitantly

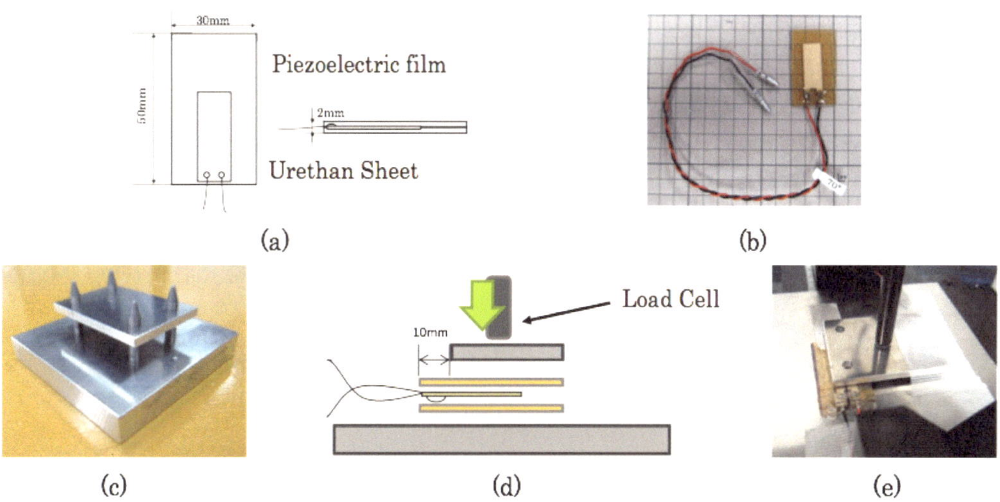

Figure 3. Load test: (c) fixture, (d) schematic diagram, and (e) load test.

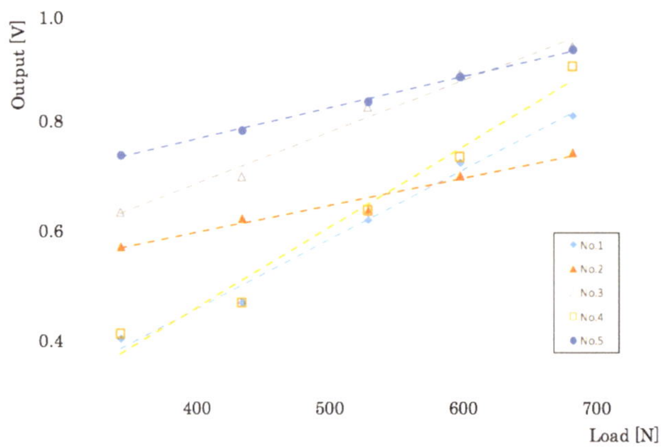

Figure 4. Relation between output voltage and load of sensors.

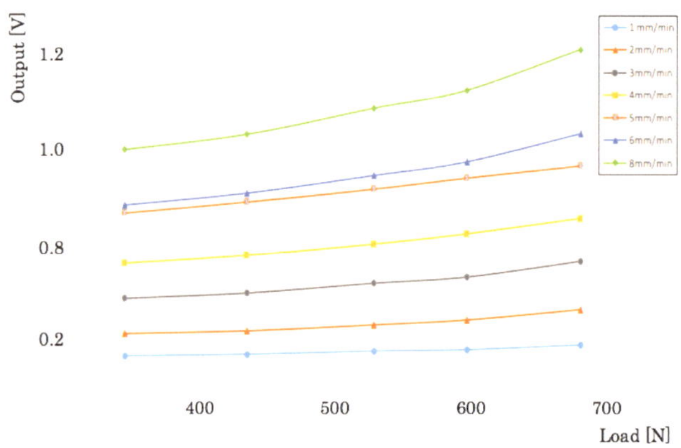

Figure 5. Relation between output voltage and load speed.

with the load until the peak for the maximum load. Subsequently, reverse voltage appears during a slight time as a steady state for removing the load cell from the device. The output characteristics were calculated using our prototype sensors obtained from this peak voltage obtained for this test. The results address the linear relation between the sensors and the test load patterns, although the gradients differ among sensors. We consider that the output voltage increases concomitantly with the weight of a person.

Figure 5 presents results of changing test speeds from 1 to 8 mm/min step by 1 mm/min. The output voltage increases concomitantly with speed changes that are similar characteristics that resemble the load-test results presented earlier.

Figure 6 depicts characteristics of the side and orientation of the sensors. Herein, four patterns were evaluated: a top/longitudinal side, a bottom/longitudinal side, a top/lateral side, and a bottom/lateral side. The top and bottom sides are defined by rivets of the piezoelectric film.

http://dx.doi.org/10.5772/intechopen.77724

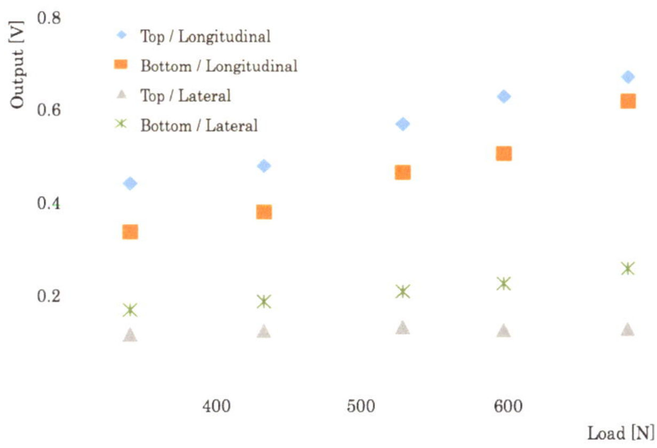

Figure 6. Relation between output voltage and load of side and orientation.

The results are depicted in **Figure 6**. The output voltage of the longitudinal side is 3.12 times higher than that of the lateral side. This directivity characteristic is reflected in the sensor installation.

6. Experimental results

6.1. Parameters

For evaluation of our developed sensor system, a similar experimental environment was set as a clinical site. Herein, the number of subjects is 10 persons: Subjects A–J. As attribute information, the body weights of the subjects are from 50 to 80 kg. They repeated the behavior sequences of seven patterns for five times. Therefore, 35 pattern datasets were obtained in each subject. Each behavior is switched in 20 s intervals. The sampling rate is set for capturing signals to 50 Hz. Herein, the number of subjects of our former study [20] was merely three persons.

Table 1 denotes optimized parameters used for SOMs and CPNs. These values were determined based on our former study [13] and the literature by Hosokawa et al. [32]. As an evaluation method, leave-one-out cross validation (LOOCV) [28] was used. For this experiment, four datasets and the remaining one dataset were used, respectively, for training and testing. Therefore, our method was evaluated for five combinational patterns in each subject.

Methods	I	J	P	Q	S	$\alpha(0)$	$\beta(0)$	O [epoch]
SOMs	10	—	10	—	16	0.5	—	1,000,000
CPNs	10	7	20	20	16	0.5	0.8	100,000

Table 1. Setting values of parameters on SOM and CPN for this experiment.

6.2. Measured signals

Figure 7 depicts time-series output signals obtained from the nine sensors. The vertical and horizontal axes depict the voltage output and transition time, respectively. The dashed lines depict boundaries of each behavior pattern. Output signal patterns are different in each subject.

For pad sensors, the voltage output range on S1 through S4 was expanded, especially in face-up, right, and left sleeping behavior patterns. Although salient voltage output included some noise, they were generated from S5 and S6. The voltage output on S3 became high for longitudinal sitting. The body weight was concentrated to the hip area according to the upper body rising behavior. For terminal sitting, the voltage output was generated from S5 and S6 and no voltage

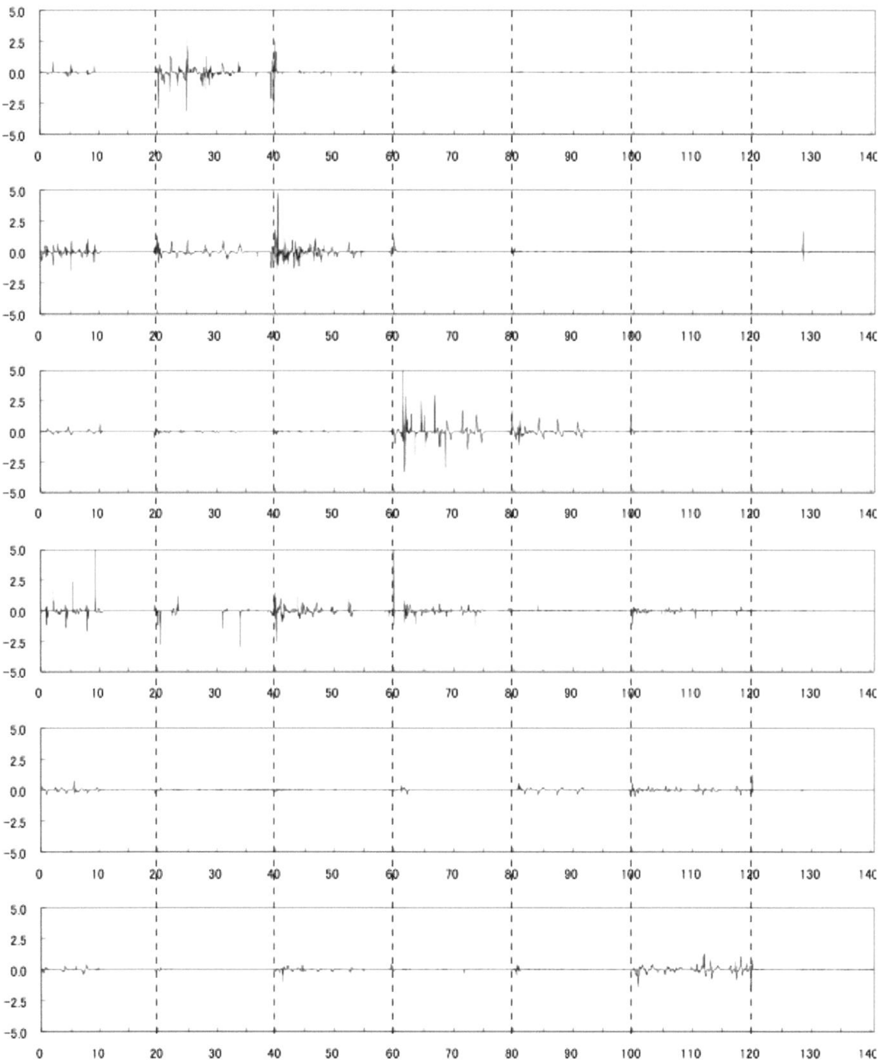

Figure 7. Time-series changes of sensor output in each behavior pattern (Subject A, first trial).

output was generated from S1, S2, and S3. Moreover, S4, which is located near S5, gave voltage output slightly. After leaving from the bed, no voltage output was generated from any sensor.

For the pillow sensor, the Z axis signal output of the upper and lower directions was greater than that of either the X or Y axis. The output became small during longitudinal sitting after body raising, which was similar tendency to that of lateral sitting. For terminal sitting, the sensor output expanded because the subject attempted to stand up at the bed terminal. This widening of output results from the behavior by which the subject tried to grip the safety handrail by hands. The weight shift at the bed terminal produced severe shaking all over the bed compared with that occurring around the center.

Signal outputs from the bolt sensor were presented only by the terminal sitting. The bolt sensor captures slight strain with the ± 0.1 width of output signals. The signal-noise rate was high compared with other two sensors. This is the dataset of Subject A at the first trial. Output signal patterns and their intensity differ in each subject and each trial. Therefore, it is solely a challenging task for setting thresholds of various types used for noise removal or for recognition solely from original signals. In contrast, our proposed sensor system can automatically remove noise using a machine-learning-based method.

6.3. Noise removal results

Using SOMs, our method removed noise patterns not only for improving recognition accuracy, but also for reducing calculation cost. The unit numbers were sorted according to the maximum of the bursting iterations from the left to the right. Most voltage output remained near the offset value because no voltage output when no load change was given without bending.

Figure 8 depicts signal output patterns of the X axis from the pillow sensor during the first 70 s period after removing sixth-unit mapped signals. As the whole tendency, noise signals around $\pm 0.1\left[m/s^2\right]$ were holistically removed. However, the signals around 60 s with higher amplitude were also removed, while those with smaller amplitude were remained.

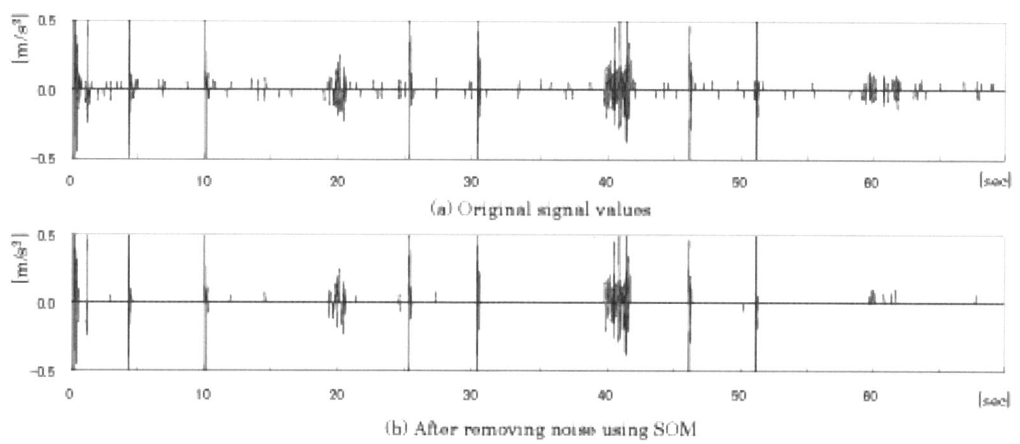

Figure 8. Noise removing result using SOM (pillow sensor, X axis).

6.4. Recognition results

For our method, category maps are used for classifiers and recognizers. **Figure 9** depicts category maps for Subjects A, B, and C. The numbers from 1 to 7 on the category maps correspond to respective behaviors that were labeled through learning. The label distributions differ among subjects. Respective category maps created clusters of similar labels for similar behaviors.

The boundaries were divided with units on the map for thick lines that depict segments of sleeping, sitting, and leaving from the bed as global changes of behavior patterns. Similar behavior patterns created recognizable behavior regions. Units, which are assigned to the seventh label as left from the bed, are distributed in the cluster of units labeled sitting. Dash marks were inserted in the category maps to the labels in each subject. Our method created a classifier in each subject through learning on CPNs. Moreover, our method enables visual verification of the relations among datasets.

Subsequently, our method was evaluated for recognizing bed-leaving behaviors from matched labels that correspond to burst unit labels after presenting test datasets to the category maps. **Table 2** portrays a result of recognition accuracies with five combination patterns evaluated as LOOCV. The recognition accuracies were calculated as a ratio of the number of ground truth (GT) labels and the number of category labels in each behavior pattern that reached the maximum of responses during 20 s.

The recognition accuracies of Subjects B and E were, respectively, 88.1 and 59.5% as the highest and the lowest. The recognition accuracies of leaving and lateral sitting were 95.0 and 53.3%, respectively. The recognition accuracies of longitudinal sitting and terminal sitting, which are the most important positions for bed-leaving detection, reached 83.3 and 98.3%, respectively. The mean recognition accuracy for 10 subjects was 75.5%. The lateral sitting as a status that can accommodate diversity among individuals. The recognition accuracy of Subject B for lateral sitting was 16.7%. For sleeping, the recognition accuracy of face-up sleeping was higher than that of right or left sleeping. The body weight gathered to the two sensors on the left or right

Figure 9. Category maps of Subjects A–C. The grids correspond to the units on the mapping layer of CPNs. Labels 1–7 denote face-up sleeping, right sleeping left sleeping, longitudinal sitting, lateral sitting, terminal sitting, and left the bed.

Subject	Face-up	Right	Left	Longitudinal	Lateral	Terminal	Left the bed	Average
A	100	83.3	83.3	100	33.3	100	100	85.7
B	100	100	100	100	16.7	100	100	88.1
C	0	66.7	100	100	83.3	100	100	78.6
D	66.7	50.0	33.3	100	66.7	100	100	73.8
E	50.0	83.3	16.7	50.0	33.3	83.3	100	59.5
F	50.0	83.3	66.7	100	66.7	100	100	81.0
G	66.7	66.7	50.0	66.7	50.0	100	83.3	69.0
H	66.7	33.3	83.3	66.7	66.7	100	100	73.8
I	66.7	50.0	66.7	66.7	66.7	100	100	73.8
J	100	66.7	33.3	83.3	50.0	100	66.7	71.4
Ave.	66.7	68.3	63.3	83.3	53.3	98.3	95.0	75.5

Table 2. Recognition accuracies of respective subjects and positions for seven patterns (%).

sides in sleeping, whereas the load from their upper body was distributed on S1–S4 at face-up sleeping.

The performance was compared with a single use of pad sensors, a pillow sensor, and a bolt sensor. **Table 3** presents the recognition accuracies of three sensor patterns. The recognition accuracy of the pad sensors as the highest performance was 75.0%. Comparison with our proposed multiple sensor system denoted that the performance difference was 0.5%. The recognition accuracies using a pillow sensor and a bolt sensor as a single sensor system were 58.3 and 33.3%, respectively. During the use of setting priority to simplicity, to use each single sensor is possible except for the bolt sensor. Our method for a multiple sensor system is useful for situations favoring performance over simplicity.

6.5. Discussion

Table 4 portrays the confusion matrix (CM) for analyzing recognition details. For this matrix, the number of correct images is assigned to the diagonal cells that are marked as boiled numbers. Other cells refer the number of incorrect images and its category name depicted in the column. The maximum numbers of incorrect categories are marked as underlined.

Sub.	Face-up	Right	Left	Longitudinal	Lateral	Terminal	Left the bed	Average
Pad sensors	35.0	81.7	75.0	90.0	55.0	96.7	91.7	75.0
Accuracies	68.0	34.0	46.0	52.0	46.0	74.0	88.0	58.3
Bolt sensor	33.3	0	11.1	0	22.2	100	66.7	33.3

Table 3. Recognition accuracies used for stand-alone sensor (%).

Position	Face-up	Right	Left	Longitudinal	Lateral	Terminal	Left the bed
Face-up	<u>40</u>	3	15	2	0	0	0
Right	5	<u>41</u>	6	2	0	6	0
Left	9	12	<u>38</u>	0	0	1	0
Longitudinal	0	0	1	<u>50</u>	7	2	0
Lateral	0	0	0	20	<u>32</u>	5	3
Terminal	0	0	0	1	0	<u>59</u>	0
Left the bed	0	0	0	3	0	0	<u>57</u>

Underlines show the numbers of correct data.

Table 4. Confusion matrix of respective positions for seven patterns.

The recognition accuracy of lateral sitting was the second lowest. The number of correct recognition datasets was 32 sets. For this behavior pattern, 20, 5, and 3 datasets are falsely recognized as longitudinal sitting, terminal sitting, and left the bed, respectively. The behavior patterns from longitudinal sitting to lateral sitting had a large gap because they turned their body about 90° on the bed. For both sitting positions, the load provided by the pad sensors was concentrated to the hip at the bed center. The false recognition in both behavior patterns was occurred from this feature. For lateral sitting, the legs load was gathered to the bed terminal as an exit. The false recognition was occurred from unclear boundaries between lateral sitting and terminal sitting. Since no dataset of lateral sitting was falsely recognized as sleeping or bed leaving, fatal errors of bed-leaving prediction can be avoided in this status transition.

Twelve and nine datasets of left sleeping were falsely recognized as right sleeping and face-up sleeping, respectively, as shown in the fourth line of **Table 4**. False recognition of left sleeping remained for sleeping except for one dataset. For face-up and right sleeping, four datasets were falsely recognized as longitudinal sitting. The results from the load to be gathered S3 merely depended on a subject, although the load on right sleeping was gathered to S1 and S3 normally.

The recognition was examined from three patterns: sleeping, sitting, and leaving. For bed-leaving prediction, these are basic behavior patterns. **Table 5** presents the CM with respective recognition accuracies. The recognition accuracies of sleeping, sitting, and leaving improved to 93.9, 97.8, and 95.0%, respectively. The mean recognition accuracy improved to 95.7%. Our system provides sufficient recognition accuracy, especially for recognition of standing behavior patterns.

Comparison with the bed-leaving sensor system that used strain gauges in actuators proposed by Hatsukari et al. [15] revealed that the recognition accuracies of their method were 87.7% for longitudinal sitting and 98.1% for terminal sitting. Although the experimental environment and the number of subjects differ in the results, the recognition accuracies of terminal sitting and longitudinal sitting of our method are, respectively, 0.2% higher and 4.4% lower than their method. Their method remains to three target patterns: longitudinal sitting, terminal sitting, and leaving. Moreover, their method is necessary for setting a subject body weight in advance. The calibration procedure differs from our system.

	Sleeping	Standing	Left	Acc. (%)
Sleeping	169	11	0	93.9
Standing	1	176	3	97.8
Left	0	3	57	95.0

Table 5. Confusion matrix with respective recognition accuracies for three patterns.

7. Conclusion

This chapter presented a multiple smart sensor system for predicting human behavior patterns that occur when a person leaves a bed. The proposed system was designed for early prediction of bed-leaving behavior patterns with regarded low privacy and high QoL. The behavior pattern prediction and recognition method was developed using machine-learning algorithms of two types from signals obtained using the sensors. Our system is applicable to an actual environment as a novel sensor system that does not restrict patients.

For future work, steady detection will be achieved to expand the range of applications and thereby increase the number of subjects. Moreover, our system must be applied to care facilities or single senior's homes for security and safety observation that simultaneously maintains QoL and privacy.

Author details

Hirokazu Madokoro

Address all correspondence to: madokoro@akita-pu.ac.jp

Department of Intelligent Mechatronics, Faculty of Systems Science and Technology, Akita Prefectural University, Japan

References

[1] Inokuchi K. The labor shortage in care workplace and employment intention of the young. Proceedings of Kanazawa University Graduate School Human and Socio-Environmental Studies. 2008;**15**:69-84

[2] Yamada R, Takashima M, Sato Y, Ito W, Ito T, Asanuma Y. Evaluation and prevention of inpatient falls: A study using a classification system based on situational criteria. Proceedings of Akita University School of Health Sciences. 2010;**18**(2):144-150

[3] Mitadera Y, Akazawa K. Analysis of incidents occurring in long-term care insurance facilities. Bulletin of Social Medicine. 2013;**30**(2):123-130

[4] Imaizumi K, Iwakami Y, Yamashita K. Availability of monitoring system for supporting healthcare of elderly people. Japanese Journal of Applied IT Healthcare. 2010;**5**(1):63-64

[5] Matsuda H, Yamaguchi A, Arakawa T. Monitoring system of living activities for elderly people. National Technical Report. 2003;**82**:4-8

[6] Tatsumi T, Kanemoto K, Yagi N. How to use efficient Tentomushi—Considering the length of stride and height for protecting false operation. Proceedings of Japanese Nursing Association. 2007;**38**:144-146

[7] Kondo S, Kamiya C, Miyamoto H, Toriyama Y, Mimura E, Tsuchida F. Availability of sensor mats to detect leaving for protection of falling accidents from bed. Journal of the Japanese Association of Rural Medicine. 2006;**55**(3):245

[8] Haruyama K, Tanaka K, Kobayashi S, Yasuoka K, Uchibori A, Oka M. Development of getting up detection and report device using power line communication and mat-sensor. Transactions of the Institute of Electric Engineers of Japan. 2006;**126**(11):1507-1513

[9] Seki H, Hori Y. Detection of abnormal action using image sequence for monitoring system of aged people. Transactions of the Institute of Electric Engineers of Japan. 2002;**122**(2):1-7

[10] Okada S, Fujiwara Y, Matsuura H, Yasuda M, Sugai K, Makikawa M, Iida T. Heart beat monitoring during sleeping using an accelerometer set inside the coverlet. Transactions of the Japanese Society for Medical and Biological Engineering. 2003;**41**(4):493-497

[11] Measurement Specialties Inc. Datasheet of DT Series Elements with Lead Attached, Rev. 1; 2009

[12] Iomoto K, Miyazaki R, Hasegawa T, Yonei Y. A trial of evaluation for quality of sleep using a three dimension acceleration sensor. The Science and Engineering Review of Doshisha University. 2010;**51**(1):28-36

[13] Tsukada M, Utsumi Y, Madokoro H, Sato K. Unsupervised feature selection and category classification for a vision-based mobile robot. IEICE Transactions on Information and Systems. 2011;**E94-D**(1):127-136

[14] Motegi M, Matsumura N, Yamada T, Muto N, Kanamaru N, Shimokura K, Abe K, Morita Y, Katsunishi K. Analyzing rising patterns of patients to prevent bed-related falls (second report). The Journal of Japan Society for Health Care Management. 2011;**12**(1):25-29

[15] Hatsukari T, Shiino T, Murai S. The reduction of tumbling and falling accidents based on a built-in patient alert system in the hospital bed. Journal of Science of Labour. 2012;**88**(3): 94-102

[16] Shimoi N, Madokoro H. Flight stabilization of micro flying robot for rescue scenarios. International Journal of Automation, Robotics and Autonomous Systems. 2012;**12**(1):11-15

[17] Shimoi N, Cuadra CH, Madokoro H, Saijo M. Simple smart piezoelectric bolt sensor for structural monitoring of bridges. International Journal of Instrumentation Science. 2012; **1**(5):78-83

[18] Shimoi N, Madokoro H, Xu L. Piezoelectric vibration measuring sensor and accelerometer used for bed monitoring system. Transactions of the Japan Society of Mechanical Engineers. 2014;**80**(812):1-14

[19] Madokoro H, Shimoi N, Sato K. Bed-leaving detection using piezoelectric unrestrained sensors and its measurement system regarding QOL. Nursing and Health. 2013;**1**(2):36-45

[20] Madokoro H, Shimoi N, Sato K. Prediction of bed-leaving behaviors using piezoelectric non-restraining sensors. Journal of Sensors and Sensor Systems. 2013;**2**(1):27-34

[21] Madokoro H, Shimoi N, Sato K. Development of non-restraining and QOL sensor systems for bed-leaving prediction. IEICE Transactions on Information and Systems. 2013;**J96–D**(12):3055-3067

[22] Madokoro H, Shimoi N, Sato K, Xu L. Pattern recognition of movements on bed aimed at prediction of bed-leaving behaviors: Time-series feature learning using Elman-type feedback counter propagation networks. International Journal of Instrumentation Science. 2015;**51**(8):528-534

[23] Madokoro H, Shimoi N, Sato K. Unrestrained multiple-sensor system for bed-leaving detection and prediction. Nursing and Health. 2015;**3**(3):58-68

[24] McQueen J. Some methods for classification and analysis of multivariate observations. In: Proceedings of Fifth Berkeley Symposium on Mathematical Statistics and Probability. 1967. pp. 281-297

[25] Nielsen H. Counter propagation networks. In: Proceedings of IEEE First International Conference on Neural Networks. 1987

[26] Barlow HB. Unsupervised learning. Neural Computation. 1989;**30**(1):295-311

[27] Kohonen T. Self-Organizing Maps. Springer Series in Information Sciences; 1995

[28] Kohavi R. A study of cross-validation and bootstrap for accuracy estimation and model selection. Proceedings of the Fourteenth International Joint Conference on Artificial Intelligence. 1995;**2**(12):1137-1143

[29] Terashima M, Shiratani F, Yamamoto K. Unsupervised cluster segmentation method using data density histogram on self-organizing feature map. Trans. Institute of Electronics, Information and Communication Engineers. 1996;**J79-D-2**(7):1280-1290

[30] Doya K. What are the computations of the cerebellum, the basal ganglia, and the cerebral cortex. Neural Networks. 1999;**12**:961-974

[31] Vesanto J, Alhoniemi E. Clustering of the self-organizing map. IEEE Transactions on Neural Networks. 2000;**11**(3):586-600

[32] Hosokawa M, Ito Y, Hoshi T. Extraction of urban characteristics using polarimetric SAR data and self-organizing map. Trans. Institute of Electronics, Information and Communication Engineers B. Japanese Journal. 2001;**J84-B**(6):1043-1051

[33] Yamakawa T, Tokutaka H, Fujimura K. Application Examples of Self-Organizing Maps. Kaibundo Pub.; 2002

Chapter 7

Piezoelectric Materials for Medical Applications

Melodie Chen-Glasser, Panpan Li, Jeongjae Ryu and Seungbum Hong

Additional information is available at the end of the chapter

http://dx.doi.org/10.5772/intechopen.76963

Abstract

This chapter describes the history and development strategy of piezoelectric materials for medical applications. It covers the piezoelectric properties of materials found inside the human body including blood vessels, skin, and bones as well as how the piezoelectricity innate in those materials aids in disease treatment. It also covers piezoelectric materials and their use in medical implants by explaining how piezoelectric materials can be used as sensors and can emulate natural materials. Finally, the possibility of using piezoelectric materials to design medical equipment and how current models can be improved by further research is explored. This review is intended to provide greater understanding of how important piezoelectricity is to the medical industry by describing the challenges and opportunities regarding its future development.

Keywords: piezoelectric materials, biotechnology, biomedical applications and devices, vital signs, sensors, cell regeneration

1. Introduction

Piezoelectricity is a quality of material asymmetry that leads to the conversion of electric signals into physical deformation and conversely physical deformation into electric signal. An applied pressure causes movement of the dipole moment within the material, and a flow of charges if crystals are aligned [1]. This makes piezoelectricity useful for a variety of industry purposes, particularly those related to vibrational generation and actuation. Commercialized applications for piezoelectricity include timekeeping using quartz resonance, microphones, radio antenna oscillators, speakers, hydrophones, and fuel injection [2, 3]. More experimental technology includes energy harvesting and electronic sensing [2]. The most commonly used ceramic piezoelectric material is lead zirconium titanate (PZT), because its physical properties can be tailored by composition, it has a high piezoelectric coefficient, and it is cheap to manufacture [4]. The most common piezoelectric polymer, used for its large strain value, is polyvinylidene fluoride (PVDF) and its copolymers, such as P(VDF-TrFE) [5–8]. A wide variety of composites and nanostructure materials have also been developed and can be fabricated as thin films, discs, or stacked sheets [2, 3, 9–13].

In the case of biomedical engineering, many conventional means of using piezoelectric devices are not applicable because of the structure of biological systems. Issues such as size limitations, biological compatibility, and flexibility have led to investigation into polymer, composite, nanostructured, and lead-free piezoelectric materials. One way to develop biomedical devices is to look at the piezoelectric structures inside the body and how they can be emulated to develop piezoelectric medical technology. In the first section of this book, we discuss piezoelectric materials present in the body. Then we describe how piezoelectric materials can be used for diagnosing illnesses and providing medical treatment. Our purpose is to inform the readers of challenges and different approaches applicable to developing a wide variety of medical technology.

2. Biological piezoelectric materials

There are many reviews which cover subsections of biological piezoelectric materials; these reviews explain topics such as piezoelectricity in bone [14] or biopolymers [15]. However, we seek to present a broader overview of the topic and how it can be used to develop technology. Much of the original work on discovering piezoelectricity in the body was done by Eiichi Fukada [15–18]. His work showed the presence of piezoelectricity in bone, aorta, muscles, tendons, and intestines [15–18]. Since that time, many further studies have contributed to the overall knowledge of the body's piezoelectric characteristics, their origins, and how they can be applied in medical science.

The organic piezoelectric effects in the human body are attributed to the lack of symmetry in most biological molecules, which may make piezoelectricity a fundamental biological property [19]. In particular, proteins seem to drive the piezoelectric qualities of most organs. The basic building blocks of proteins within the human body are amino acids. These make up molecules such as collagen, keratin, and elastin which are highly prevalent in the organs examined by Fukada and other researchers [15–18]. Amino acids in pure form have their own piezoelectric properties due to the presence of dipoles derived from the polar side groups seen in **Figure 1**. It is the reorientation and change in dipole moments in biological macromolecules under stress that gives them piezoelectric properties [20, 21]. At least 15 amino acids, mostly the "L" form, exhibit piezoelectric properties; however, γ-glycine and DL-alanine are the strongest amino acid piezoelectrics [22]. Most racemic, or DL mixtures of amino acids do not show piezoelectric properties because their crystal forms are centrosymmetric [23].

http://dx.doi.org/10.5772/intechopen.76963

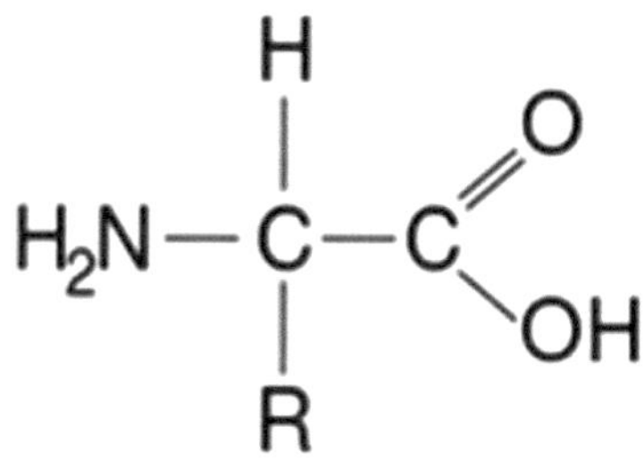

Figure 1. The general structure of amino acids. Reprinted and altered from Ref. [24].

Other biological piezoelectric materials include polymeric L-lactic acid, DNA, and the M13 bacteriophage [25–27]. Like amino acids, the piezoelectric properties of lactic acid come from the carbon–oxygen double bond [25]. DNA's piezoelectric properties originate from internal rotation of the dipoles created by phosphate groups; however, they were primarily observed at lower water content, which makes the bonds holding the DNA helix together weaker [27]. This demonstrates the importance of bonding, structure, and experimental conditions when determining piezoelectric properties. The M13 bacteriophage's piezoelectric effect is caused by extruding proteins and it can be fabricated into thin films that exhibit strengths of 7.8 pm/V [26].

Like the bacteriophage, many organs contain macromolecules which give them piezoelectric properties. Organs with piezoelectric properties can be viewed as amorphous organic material containing structured fibers which give them their piezoelectric properties [19, 28]. Often these fibrils will grow in a helix shape, preventing them from having centrosymmetric symmetry [29]. The overall strength of the piezoelectric effect will depend on the ordering, quantity or composition of these fibers. Bones and tendons have hexagonal symmetry and contain the following piezoelectric constant d_{ij} in the form of Eq. (1) [30]. In this tensor, the "i" subscript represents direction of electric field displacement and the "j" subscript represents the mechanical deformation associated with it [31].

$$d_{ij} = \begin{pmatrix} 0 & 0 & 0 & d_{14} & d_{15} & 0 \\ 0 & 0 & 0 & d_{15} & -d_{14} & 0 \\ d_{31} & d_{32} & d_{33} & 0 & 0 & 0 \end{pmatrix} \tag{1}$$

Molecular structure within the organ changes the organ's overall piezoelectric nature. For example, examination of the epidermis, horny layer, and dermis of the skin revealed that each layer had its own piezoelectric coefficient, the highest being the horny layer. The dermis had a less ordered collagen layer; the horny layer had parallel keratin filaments, and the epidermis α helical keratin tonofibrils [28]. The structure of the keratin horny layer simplified its ability to produce piezoelectric tensors, giving them the form of Eq. (2). The values of piezoelectric coefficients varied based on temperature; however, the highest were seen in the horny layer, on the order of 0.1–0.2 pC/N. The lack of consistency in these measurements is due to the variety in how the molecules were ordered in each sample [28].

$$d_{ij} = \begin{pmatrix} 0 & 0 & 0 & d_{14} & 0 & 0 \\ 0 & 0 & 0 & 0 & -d_{14} & 0 \\ 0 & 0 & 0 & 0 & 0 & 0 \end{pmatrix} \tag{2}$$

Similarly, piezoresponse force measurements (PFM) studies of collagen proved that collagen is the main source of piezoelectricity in the bone and reveal different ordering of collagen fibers results in different piezoresponses, as seen in **Figure 2** [32]. In collagen, there are alternating sections of overlap and gap regions. The collagen fibers are arranged in a staggered way that result in the gap region having one less microfiber. In addition, the molecules in the gap region have less uniform symmetry, and therefore that region does not have as high of a piezoresponse [32]. These two studies indicate the piezoelectric response is not merely dependent on the molecular structure, but the structure of the entire organ. **Table 1** gives a description of organs with tested piezoelectric properties and their attributed molecule.

Despite many measurements, it is sometimes difficult for the scientific community to come to a consensus on the exact nature and relevance of *in situ* piezoelectric characteristics. For example, in the case of bone, two groups found contradicting results on the dependency of piezoelectricity in terms of hydration [14, 37]. Some studies on the aorta indicate that it has piezoelectric properties, though results were varied. Two studies, taken over forty years apart showed different orders of magnitude for the studied properties [17, 38]. A lab attempting to verify either of these studies found that there was no piezoelectric response from the aorta [39].

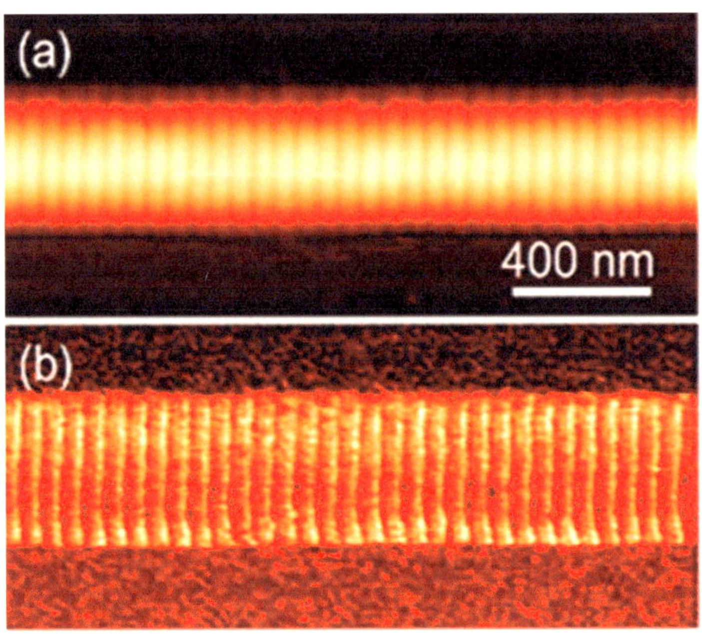

Figure 2. The images show (a) the topology of the collagen and (b) the piezoresponse force microscopy (PFM) image where the collagen can be distinguished from the surrounding tissues and how the gap and overlap regions differ in piezoelectric response. Reprinted from Minary-Jolandan and Yu [32] with permission from ACS Publications.

Organ	Piezoelectric molecule
Muscle	Actin and myosin [18]
Hair	Keratin [16]
Bone	Collagen [32]
Tendon	Collagen [33]
Lung tissue	Elastin [34]
Skin (dermis)	Collagen [28]
Skin (horny layer and epidermis)	Keratin [28]
Breast tissue	Collagen [35]
Outer hair cell	Prestin [36]

Table 1. Tissues with piezoelectric properties and driving source of piezoelectricity.

Historically, piezoelectric potentials were thought to explain Wolff's Law, the fact that bone is strong in areas that are subject to greater amounts of stress [40]. However, later research proved streaming potentials, fluid and ions driven by mechanical loading, may have a greater impact in determining bone properties [41]. However, Ahn *et al.* suggest that piezoelectricity could generate charges that affect the screening potential and the two work in conjunction to promote bone development, a concept which requires experimental testing to verify [40]. Furthermore, the generation of electric fields has been shown to increase bone healing during fracture [42, 43].

Despite the variety of results concerning piezoelectric qualities of the body, they do help in understanding the body's mechanics and how we can develop solutions for human problems. Even if the exact purpose for piezoelectric properties in the body is not known, they still can be used for developing biomedical solutions on both microscopic and macroscopic levels. For example, knowing that amino acids and macromolecules composed of them have piezoelectric properties has inspired the use of biomaterials for human sensors [44]. Using peptides to build piezoelectric sensors eliminates the need for developing other biocompatible materials. For example, the knowledge of previously mentioned virus, M13, led to the alignment of its phages into nanopillars for enhanced piezoelectric properties [45]. The outer hair cell is another structure that piezoelectric properties can be attributed to. Disruption of the cell's electrical potential alters its length; conversely, compression of the cell alters its membrane potential [46]. The motions of the outer hair cell alter how the organ of Corti vibrates, and changes how the inner hairs receive stimulation [36]. Recently, the development of a piezoelectric cochlear implant to mimic the conversion of sound vibration into an electrical signal has been undertaken and will be covered in a later section of this review [47]. Biological structures can serve as examples for the development of piezoelectric structures and biocompatible piezoelectric materials.

In addition, the knowledge of piezoelectric properties can help in disease detection or injury analysis. With the knowledge that piezoelectric tissue properties are determined by proteins, diseases that affect the amount or distribution of these proteins can be detected by piezoelectric sensors. One group proposed that the electromechanical coupling factor, controlled by collagen, could aid in detecting breast cancer [35]. A similar idea was presented for the detection of atherosclerosis in the aorta, however as mentioned in a prior paragraph the validity of the aorta's piezoelectric nature is still under debate [48]. In this paper, they claimed the PFM amplitude increased as a function of advancing atherosclerosis and could help with early detection of the disease.

Finally, once the effect of piezoelectricity on the body have been studied, piezoelectric materials can be used to promote disease healing. Though the exact reason for piezoelectric qualities have not been fully discovered, studies into bone related injuries have revealed that induced electrical fields can accelerate bone repair and promote the growth of neurons [49, 50]. Because of this, increasing the piezoelectric properties of a synthetic bone material has potential to increase the speed of osteoconduction and subsequently bone repair [51]. Lead free ceramics can be used in conjunction with synthetic bone; however, these materials have problems with ion diffusion which can be controlled by embedding in a ceramic or polymer matrix [50]. In terms of regenerating damaged bone or cartilage, a piezoelectric scaffold may provide the necessary stimulation for cell regrowth, and diminish the need for other growth factors [43]. Typically, scaffolds are made out of polymers, such as PVDF, and can also promote the growth of neurons and wound healing [50].

3. Piezoelectric medical devices

Many biomedical piezoelectric applications exceed the aforementioned purposes of mimicking or employing biological piezoelectric phenomena. In some cases, the choice of material depends mostly on the strength of the piezoelectric effect and the cost of the material. PZT (lead zirconium titanate) and quartz are common piezoelectric materials used in industry. PZT is cheaper, has higher piezoelectric coupling coefficients, and can be manipulated by changing the composition. Quartz, however, is more stable and has consistent properties over a broader temperature range [4]. Developing implants or technology involving direct human contact has more constraints. Ceramics, like quartz, barium titanate, and potassium sodium niobate, are more biocompatible because they do not contain lead [50]. In addition, many biomedical devices require higher flexibility than ceramics can provide, due to the dynamic nature of human motion. Biocompatible polymers include most biological materials and PVDF copolymers. So far, polymer applications of PVDF have included, but are not limited to, biomechanical energy harvesting systems, sensors, and wound scaffolds [50, 52]. The piezoelectric coefficient of the beta phase of PVDF is listed in Eq. (3) [53].

$$d_{ij} = \begin{pmatrix} 0 & 0 & 0 & 0 & d_{15} & 0 \\ 0 & 0 & 0 & d_{24} & 0 & 0 \\ d_{31} & d_{32} & d_{33} & 0 & 0 & 0 \end{pmatrix} \tag{3}$$

3.1. Piezoelectric sensors

Piezoelectric materials can be employed in monitoring many bodily signals because they convert mechanical energy into an electrical signal. They are especially applicable to monitoring dynamic pressure changes; many human vital signs consist of rhythmic activities like the heartbeat or breathing. Lower pressure systems from (1 Pa-10 kPa) include sound waves and tactile sensing. In the higher end of that range are intraocular pressure and cranial pressure. Higher-pressure systems (10 kPa–100 kPa) correspond to blood pressure measurements and some bodily movements. Piezoelectric sensors can be tailored by structure or material to match the pressure range of the desired quality [54]. Implanted or wearable medical sensors have greater applicability, as the Internet of Things becomes more fully developed. A medical professional or computer algorithm can monitor a patient for early warning signs that may have been missed between scheduled check-ups through their implanted device [55]. **Table 2** lists some literature studies of piezoelectric sensors and their tested applications.

The variety of applications for piezoelectric sensors in the biomedical industry is promising, however much of this technology is still in the research and development phase. Before reaching the market, these devices need to have scalable manufacturing and guaranteed quality for every device [52].

3.1.1. Developing synthetic skin

A specific application for piezoelectric pressure sensing is synthetic skin. As a bare minimum, synthetic skin should provide the magnitude of contact force and approximate location of

Material	Applications	Device characteristics	Refs.
Prawn cell	Wrist pulse	100 Hz–10 MHz range	[56]
PVDF	Human voice detection Hand motion Breathing rate	50–1000 Hz range	[57]
$(Na_{0.5},K_{0.5})NbO_3$ (NKN) thin film	Cardio mechanical electric sensor	10 Hz resonance	[58]
PVDF	Wrist pulse Measuring peripheral arterial pressure pulse		[59]
AlN	Heart and respiration patterns for sleep apnea	Tested over 0.1–10 Hz	[60]
PVDF	Heartbeat and respiration detection	Tested 0.1–2 Hz	[61]
Fish gelatin	Joint movement Human vocal cord movement Radial artery pulses	d_{33}–20 pm/V Stability over 108,000 cycles	[62]
PZT	Eye fatigue via eyelid motion		[63]
Poly-L-lactic acid	Lung pressure Eye pressure Brain pressure	Biodegradable Stability over 108,000 cycles	[64]
Piezoelectric ceramic	Vision correction	Force sensitivity 0.1×10^{-2} N to 5×10^{-2} N 0.01–5 Hz	[65]
PVDF	Food detection by swallowing pattern	Limit of detection: 1 Hz Tested over 1–5 Hz	[66]

Table 2. Examples of piezoelectric sensors and their applications.

contact with the sensitivity of normal skin [53]. For humans, mechanoreceptors have a range from 3 to 400 Hz, and a spatial resolution of 1–2 mm [67]. Ideally, it would also provide information about temperature changes or humidity [68]. Human skin itself acts as a vibrational sensor; it is structured to amplify tactile stimulation [69]. Piezoelectric force transducers offer a solution to quantifying and locating contact forces [53]. The use of polymers for synthetic skin is popular because of their similarity in texture and flexibility to human skin [70]. Polymers can be molded to emulate human characteristics, such as fingerprints to enhance their sensitivity [69]. Processing techniques, such as electrospinning, can increase response by aligning the molecular dipoles [25]. In a similar way, using hybrid materials or structuring ceramics and polymers can yield higher piezoelectric properties [71, 72].

Though there are many materials, which can be used for this purpose, most are structured in arrays. A unit in the array will send an electrical signal describing the characteristic of the force. In prosthetics, the electric signal will arrive at a location which still can perceive tactile senses [53]. One of the problems with arrays is interference between signals, otherwise known as crosstalk. During crosstalk, neighboring units are affected by the unit undergoing force and send their own signal. This can lead to an ill-defined contact region, which can be fixed using the installation of transistors or through triangulation of the signal [53, 68].

3.1.2. Biological quartz microbalance

One other interesting application of piezoelectric sensors is the detection of disease or odor through a change in chemical composition of a sensor. The quartz microbalance is used for a variety of purposes, such as gas detection [73], composition analysis, and chirality classification [74]. It can also sense changes in liquid density or viscosity [75]. This method relies on mass changes in a coating film around the crystal. Quartz microbalances generally operate in a 5–10 MHz range; the accumulation of mass can be quantified by the Sauerbrey equation (Eq. (4)) [75, 76]. An increase in mass indicates a decrease in the frequency of quartz vibration [76]. When this mass becomes too great (>2%) this relationship becomes inaccurate, and a better approximation is needed [74]. In this type of sensor, biological molecules are imbedded or attached to piezoelectric materials. This technology can also be used for detection of bacteria and biomolecules.

$$\frac{\Delta f}{f_0} = -\frac{\Delta m}{m} \tag{4}$$

The detection of bacteria or biomolecules usually involves the incorporation of a biomolecule in an exterior film. One method of detecting glucose uses the enzyme hexokinase embedded in a polymer matrix. The glucose binds to the enzymes at a rate proportional to its concentration in solution [77]. In another glucose detection system, the frequency of the quartz was increased. The sensor was coated with dextran and Concanavalin A. The dextran preferentially binds to the glucose, therefore the presence of glucose causes the release of Concanavalin A. Glucose has a lower molecular weight, and therefore the frequency increased with its detachment. This method of glucose detection is advantageous because it does not involve the use of enzymes; however has a lower detection range [78]. The quartz microbalance may also be applicable to developing bioelectronic olfactory replacements. It has been used to detect hazardous odorants such as diacetyl, which can cause damage to the lung if inhaled, and could be used to measure other odors [79]. Unfortunately, some of the quartz microbalance equipment is bulky and requires complicated molecules as indicators. One of the olfactory biosensors is 14 mm in diameter [80]. If the synthetic nose to be used for many compounds, the size may be too large to be practical. In addition, sensors based on biomolecules, such as the glucose have problems with biological stability [78]. These problems need to be fixed before they can be viewed as commercially viable.

3.1.3. Cochlear implants

The destruction of inner ear cells results in severe hearing loss and is most commonly treated by cochlear implants. Though the current technology allows for recovery from deafness, it is incompatible with water and has very high-power requirements [81]. Piezoelectric materials can be used for creating an artificial basilar membrane (ABM). The membrane performs mechanical frequency selectivity for the cochlea. Varying physical rigidity and thickness of the basilar membrane allows it to perform its duty, and likewise piezoelectric materials can filter out frequency based on their physical properties [82]. Ceramics, such as PZT or AlN films, can be fabricated in beam or cantilever arrays with lengths corresponding with different resonance frequencies [81, 83]. Alternatively, devices based on PVDF or P(VDF-TrFE) membranes have been fabricated [47, 83–85]. The typical range of human hearing is 20 Hz–20 kHz. The fabricated PVDF membrane was able to detect signals in the 100 Hz–10 kHz range, which encompasses the range of human vocalizations [84]. Many experimental cochlear ABMs need increased sensitivity, stability, and size reduction to be practically used [83].

3.2. Beyond sensors

3.2.1. Piezoelectric surgery

In addition to creating implants, piezoelectricity can be used in a variety of medical treatments, most of which depend on the vibrational properties of the piezoelectric device. Unlike implanted devices, piezoelectric devices needed for surgery do not need to be biocompatible, because they do not come in contact with human cells. Therefore, many external devices will make use of lead zirconate titanate (PZT), as it is easier to produce [86]. The typical piezosurgical devices will consist of stacked rings which are given an applied voltage. The stacked actuator design increases the actuator efficiency because the electric field is determined by the applied voltage and the thickness (Eq. (5)) [87]. The strain is proportional to the electric field if the thickness of the actuator is decreased, a higher strain can be generated for the same amount of voltage.

$$E = \frac{V}{t} \tag{5}$$

The resulting vibration will be transduced to the tip, which is installed in such a way that it will amplify vibrations, because traditionally ceramics are more brittle and do not display much displacement [88].

In surgery, piezoelectric devices, such as the ultrasonic lancet, are used for delicate operations to preserve surrounding tissue. By controlling the micromovements of the oscillating device, damage to soft tissues can be avoided, and the separation between interfaces is easily accomplished. Alternatives to piezosurgery, such as a chisel and hammer or rotating saw are seen as more invasive, have potential to lacerate non-discriminatorily [89]. Hard tissues, such as mineralized bones are damaged by frequencies of 25–39 kHz, however neurovascular tissue is cut at frequencies higher than 50 kHz. There are no macrovibrations which may cause discomfort to the patient or disturbance of surrounding tissue [90]. The tip oscillates in a linear direction, and can span the distance of 60–200 μm [86].

The first use of piezosurgery was the dental industry, with applications like removal of implants, bone harvesting, and inferior alveolar nerve detachment [91]. Many such surgeries require working in small spaces and do not require larger incisions on the bone material. The removal of implants takes advantage of how the ultrasonic vibrations target the interfacial layer, and weaken the implant's attachment to bone. This reduces the adhesion forces and allows the implant to be removed with fewer incisions. In a similar way, the collection of graft material is another excellent use of an ultrasonic lancet. After making preliminary cuts with a saw, the ultrasonic vibrations reduce the need for chisel strikes [91]. In surgery performed on the lower jawline, protecting the inferior alveolar nerve is important to patient recovery [92]. As said previously, the use of piezosurgery prevents the damage of these nerve tissues. Another benefit in all surgeries is particle breakdown caused by ultrasonic activity, which makes visibility easier [92].

Piezosurgery has some other applications in neurosurgery and orthopedic surgery; however, it is limited in equipment fragility and associated expenses [86, 90]. The tip of the device fractures, creating the need for replacements [88]. It also takes longer to perform operations, and can damage tissue through heating. Irrigation is required to keep the area cool, and larger scale devices are used for macrosized surgeries [86].

3.2.2. *Ultrasonic dental scaling*

A piezoelectric dental scalar also has piezoelectric ceramic rings (**Figure 3**) on the inside to induce axial vibrations, and operates at ultrasonic frequencies [93]. Ultrasonic dental scalers operate in the range of 25–50 kHz, and oscillate parallel to the tooth surface over a range of 10–100 μm [94]. Its purpose is to remove accumulated biofilms from the tooth surface and for treatment of root canals. The vibrations of the tip break down the calculus (tartar) and plaque which have formed on the tooth's enamel surface. Because of the tip's quick speed, when the irrigation water passes over the scalar, micro- and nanosized bubbles form around its curve and tip [95, 96]. When these

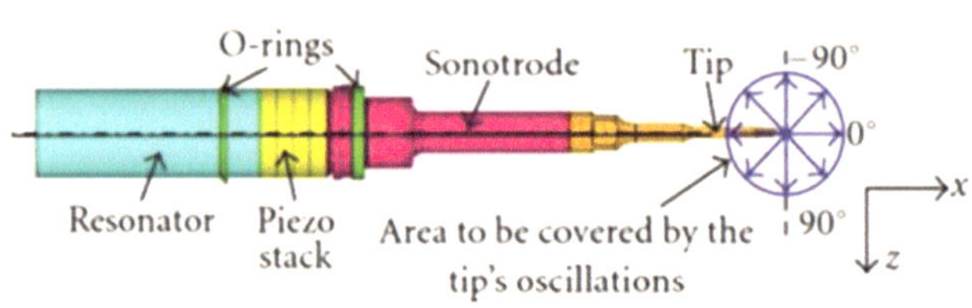

Figure 3. The stacked actuator design and other components of an ultrasonic dental scaler. Reprinted from Engelke et al. [93] with permission from Hindawi.

bubbles collapse, cavitation forces create shock waves cleaning the tooth. This adds to the ultrasonic scalar's effectiveness, and further investigation of the cavitation's effects could lead to new dental technology which further reduces scaler contact with teeth [97].

The oscillation pattern of the scaler depends on the type of tip chosen [93] and the effectiveness of the scaler varies depending on how it is used. Influencing factors can be the lateral force, tip angle, and power setting [98]. Increasing the power on the ultrasonic scalar too much will scratch a tooth's protective enamel surface, increasing the tooth's surface roughness and causing damage to the surrounding tissues. The variety of ultrasonic scalers' operating conditions demonstrates the need for research on appropriate forces needed to remove dental tartar without causing damage.

3.2.3. *Microdosing*

Microdosing is another application of piezoelectrics and has become popular because it conserves the amount of medication dispensed and can reduce discomfort by avoiding injections [99, 100]. In some cases, an injected drug can be aerosolized in order to avoid injection. In this case, piezoelectric vibrations can break the drug into fine particles which can be carried in an air stream and inhaled by the patient [100]. In the case of solids, a stacked actuator design is applicable by providing a single oscillation, rather than the consistent vibration of the previously mentioned ultrasonic devices. A glass tube is attached to the actuator and an electric signal stimulates the actuator providing a force to the tube and displacing a certain amount of the solid. Though the dispensing is very precise, it does have a minimum dosage and blockage can occur in the glass tube [101].

Stacked actuators in fluid pumps can administer small single doses or a continuous flow [102]. Fluid administration, like those for eye drops, often require small single doses [99]. A more complex form of controlled microdosing can be accomplished through a diaphragm pump [103]. This is more suited for some dosing systems such as insulin dispensing. One pump design places four chambers in series, with electrodes connecting the gate so they operate in tandem. Here the voltage controls the degree of membrane fluctuation and the phase of the material controls its direction [104]. An alternative design has parallel cylinders which are filled and emptied according to a certain sequence. The number of steps in the sequence determines the flow rate [105]. Though PZT is a popular material for biomedical pumps, polymer actuators such as PVDF-TrFE have been used as well [106].

3.2.4. *Energy harvesting*

In order to have implantable sensors within the body, they need to have a convenient source of energy. If the sensor is battery powered, future surgery will be required to extract and replace the battery. This is a current problem with pacemakers and limits the number of sensors placed after surgery. Energy harvesting through the body's movement via piezoelectricity is one way to avoid the need for battery incorporation or replacement. Energy harvesting from organs or the human body requires specific considerations, the most important being biocompatibility. Like implants, energy harvesting devices ideally should not contain hazardous chemicals, like lead, or must be sealable [107]. In the field of energy harvesting, one of the key modes of energy harvesting, a vibrating cantilever, is not as applicable to *in situ* biological energy harvesting because a vibrating cantilever has a very high resonance frequency for peak power generation [108]. Though a cantilever's resonance frequency can be changed by altering its physical characteristics, such as adding a proof mass or increasing size [109], an implant needs to be small as to avoid interference with organ function [108]. Typically, piezoelectric biomedical harvesters will be thin films that target tiny irregular vibrations caused by normal organ deformation [110]. A piezoelectric energy harvester provides an AC power source, and the most energy is gained near the resonance frequency of the film [111]. This adds another engineering constraint as biological motions usually have low natural frequencies. The human heart beats at around 39 Hz and the frequency of someone walking is around 1 Hz [108, 112]. The energy harvesting element also has to be small, as large devices may impede the normal function of human organs or cause discomfort.

Most implanted devices should have some degree of flexibility for use in the human body. Both polymer and ceramic flexible devices can be adhered to consistently moving body parts to provide a source of energy. This could include wrapping a piezoelectric film around a pulsing artery or anchoring it to an expanding diaphragm, lung, or heart [108]. The heart, or locations near it, are advantageous places to put an energy harvesting device because they could power a pacemaker. Ceramic nanoribbons are usually attached to some flexible film such as polyimide, polyethylene terephthalate, or polyethylene naphthalate. The ceramic components, made of PMN-PT, PZT, or $BaTiO_3$, are fabricated in small units and then transferred to the flexible film [110]. PVDF and PVDF-TrFE thin films can also be used to fabricate energy harvesters. These films have the advantage of being biocompatible and do not have to be transferred onto a flexible matrix [113].

In replacement joints, stacked ceramic sheets are preferred for energy harvesting. Knee surgery is a difficult process and complications can arise after surgery [112]. The replacement joint can become imbalanced, and be subject to wear, loosening, or even fracture. The presence of sensors in the replacement joint vicinity would allow doctors to study how to improve knee replacements and detect problems with greater speed. The stacked actuator is the best design for energy harvesting in implanted joints. These actuators do not need to be as flexible, because the downward force from the knee is compressive, rather than stretching [114]. Prospective locations for the actuators could be in the tibial component of the joint or in the polyethylene cartilage imitation [112, 114].

The main limitations of piezoelectric energy harvesting are low efficiency and power output. This is large concern with biomedical devices, because they often do not operate at the device's resonance frequency [115]. Another avenue of research focuses on enhancing the efficiency of energy harvesting by mechanically scraping screening charges found on the surface of piezoelectric materials [116–119].

4. Conclusion

The purpose of this book chapter has been to give an overview of piezoelectric in the biomedical industry. We have described the piezoelectric properties of biological materials and how they can be used to develop disease treatment. We also covered piezoelectric materials used in sensors, and other devices to explore the current industries which can be improved by further research. By describing these challenges, we hope to bring greater understanding of how important piezoelectricity is to the medical industry and the opportunities it has for future development.

Acknowledgements

This research was supported by the KUSTAR-KAIST Institute, KAIST, Korea and Creative Materials Discovery Program from National Research Foundation of Korea funded by the Ministry of Science and ICT (NRF-2017M3D1A1086861). The authors gratefully acknowledge Dr. Sangmin Shin for his advice and stimulating discussion.

Dr. Panpan Li was also supported by the Korea Research Fellowship Program funded by the National Research Foundation of Korea (no. 2017H1D3A1A01054478).

Author details

Melodie Chen-Glasser, Panpan Li, Jeongjae Ryu and Seungbum Hong*

*Address all correspondence to: seungbum@kaist.ac.kr

Department of Materials Science and Engineering, KAIST, Daejeon, Korea

References

[1] Nye JF. Physical Properties of Crystals: Their Representation by Tensors and Matrices. Oxford: Clarendon Press; 1985

[2] Heywang W, Lubitz K, Wersing W. Piezoelectricity: Evolution and Future of a Technology. New York, NY: Springer Science & Business Media; 2008

[3] Topolov VY, Bowen CR, Bisegna P. New aspect-ratio effect in three-component composites for piezoelectric sensor, hydrophone and energy-harvesting applications. Sensors and Actuators A. 2015;**229**:94-103

[4] Mohammadi MM. A comparison between quartz and PZT ceramic for Sensoric applications. Research Desk. 2015;**2**(4):321-325

[5] Ramadan KS, Sameoto D, Evoy S. A review of piezoelectric polymers as functional materials for electromechanical transducers. Smart Materials and Structures. 2014;**23**(3): 033001

[6] Kim D, Hong S, Hong J, Choi YY, Kim J, Park M, et al. Fabrication of vertically aligned ferroelectric polyvinylidene fluoride mesoscale rod arrays. Journal of Applied Polymer Science. 2013;**130**:3842-3848

[7] Choi YY, Sharma P, Phatak C, Gosztola DJ, Liu Y, Lee J, et al. Enhancement of local Piezoresponse in polymer ferroelectrics via nanoscale control of microstructure. ACS Nano. 2015;**9**:1809-1919

[8] Choi YY, Tong S, Ducharme S, Roelofs A, Hong S. Charge collection kinetics on ferroelectric polymer surface using charge gradient microscopy. Scientific Reports. 2016;**6**: 25087

[9] Cheng KC, Chan HLW, Choy CL, Yin Q, Luo H, Yin Z. Single crystal PMN-0.33PT/epoxy 1-3 composites for ultrasonic transducer applications. IEEE Transactions on Ultrasonics, Ferroelectrics and Frequency Control. Sep 2009;**50**(9):1177-1183

[10] Setter N, Damjanovic D, Eng L, Fox G, Gevorgian S, Hong S, et al. Ferroelectric thin films: Review of materials, properties and applications. Journal of Applied Physics. 2006;**100**: 051606

[11] Kim D, Hong S, Lee D, Roh HS, Ahn G, Kim J, et al. A spring-type piezoelectric energy harvester. RSC Advances. 2013;**3**:3194-3198

[12] Kim D, Roh HS, Kim Y, No K, Hong S. Selective current collecting design for spring-type energy harvesters. RSC Advances. 2015;**5**:10662-10666

[13] Choi YY, Yoon TG, Qaiser N, Paik H, Roh HS, Hong J, et al. Vertically aligned P(VDF-TrFE) core-shell structures on flexible pillar arrays. Scientific Reports. 2015;**5**:10728

[14] Telega JJ, Wojnar R. Piezoelectric effects in biological tissues. Journal of Theoretical And Applied Mechanics. 2002;**3**(40):723-759

[15] Fukada E. Piezoelectric properties of organic polymers. Annals of the New York Academy of Sciences. 1974;**238**:7-25

[16] Fukada E. Piezoelectricity of natural biomaterials. Ferroelectrics. 1984;**60**:285-296

[17] Fukada E, Hara K. Piezoelectric effect in blood vessel walls. Journal of the Physical Society of Japan. 1969;**26**(3):777-780

[18] Fukada E, Ueda H. Piezoelectric Effect in Muscle. Japanese Journal of Applied Physics. 1970;**9**(7):844

[19] Shamo MH, Lavine LS. Piezoelectricity as a fundamental property of biological tissues. Nature. 1967;**213**:267-269

[20] Lemanov VV, Popov SN, Pankova GA. Piezoelectric properties of crystals of some protein aminoacids and their related compounds. Physics of the Solid State. 2002;**44**(10): 1929-1935

[21] Zhou Z, Qian D, Minary-Jolandan M. Molecular mechanism of polarization and piezoelectric effect in super-twisted collagen. ACS Biomaterials Science & Engineering. 2016;**2**: 929-936

[22] Lemanov VV, Popov SN, Pankova GA. Piezoelectricity in protein amino acids. Physics of the Solid State. 2011;**53**(6):1191-1193

[23] Delfino M. A comprehensive optical second harmonic generation study of the non-centrosymmetric character of biological structures. Journal of Biological Physics. 1978;**6** (3–4):105-117

[24] Creative Commons Attribution-Share Alike 3.0 Unported. Amino Acid Zwitterions. Wikipedia; 2012. [Online]. Available: https://en.wikipedia.org/wiki/Amino_acid

[25] Sultana A, Ghosh SK, Sencadas V, Zheng T, Higgins MJ, Middya TR, et al. Human skin interactive self-powered wearable piezoelectric bio-e-skin by electrospun poly-L-lactic acid nanofibers for non-invasive physiological signal monitoring. Journal of Materials Chemistry B. 2017;**5**:7352-7359

[26] Lee BY, Zhang J, Zueger C, Chung WJ, Yoo SY, Wang E, et al. Virus-based piezoelectric energy generation. Nature Nanotechnology. 2012;**7**:351-356

[27] Zhu Y, Zhang S, Wen J. Influence of orientation on the piezoelectric properties of deoxyribonucleic acid. Ferroelectrics. 1990;**101**(1):129-139

[28] De Rossi D, Domenici C, Pastacaldi P. Piezoelectric properties of dry human skin. IEEE Transactions on Electrical Insulation. 1986;**El-21**(3):511-517

[29] Sadoc JF, Rivier N. Boerdijk-Coxeter helix and biological helices. European Physical Journal B. 1999;**12**:309-318

[30] Wojnar R. Piezoelectric phenomena in biological tissues. In: Piezoelectric Nanomaterials for Biomedical Applications. Berlin, Heidelberg: Springer; 2012. pp. 173-185

[31] Jordan TL, Ounaies Z. Piezoelectric Ceramics Characterization. NASA, Langley Research Center: Hampton, Virginia; 2008

[32] Minary-Jolandan M, Yu MF. Uncovering nanoscale electromechanical heterogeneity in the subfibrillar structure of collagen fibrils responsible for the piezoelectricity of bone. ACS Nano. 2009;**3**(7):1859-1863

[33] West CR, Bowden AE. Using tendon inherent electric properties to consistently track induced mechanical strain. Annals of Biomedical Engineering. 2012;**40**(7):1568-1574

[34] Jiang P, Yan F, Esfahani EN, Xie S, Zou D, Liu X, et al. Electromechanical coupling of Murine lung tissues probed piezoresponse force microscopy. ACS Biomaterials Science & Engineering. 2017;**3**:1827-1835

[35] Park K, Chen W, Chekmareva MA, Foran DJ, Desai JP. Electromechanical coupling factor of breast tissue as a biomarker for breast cancer. IEEE Transactions on Biomedical Engineering. 2018;**65**(1):96-103

[36] Brownell WE. The piezoelectric outer hair cell. In: Vertebrate Hair Cells. Verlag New York: Springer; 2006. pp. 313-347

[37] Reinish GB, Nowick AS. Piezoelectric properties of bone as functions of moisture content. Nature. 1975;**253**:626-627

[38] Liu Y, Zhang Y, Chow MJ, Chen QN, Li J. Biological ferroelectricity uncovered in aortic walls by piezoresponse force microscopy. Physical Review Letters. 2015;**108**:078103

[39] Lenz T, Hummel R, Katsouras I, Groen WA, Nijemeisland M, Ruemmler R, et al. Ferroelectricity and piezoelectricity in soft biological tissue: Porcine aortic walls revisited. Applied Physics Letters. 2017;**111**(13):133701

[40] Ahn AC, Grodzinsky AJ. Relevance of collagen piezoelectricity to "Wolff's Law": A critical review. Medical Engineering & Physics. 2009;**31**(7):733-741

[41] MacGinitie LA, Stanley GD, Bieber WA, Wu DD. Bone streaming potentials and currents depend on anatomical structure and loading orientation. Journal of Biomechanics. 1997; **30**(11–12):1133-1139

[42] Rodriguez R, Rangel D, Fonseca G, Gonzalez M, Vargas S. Piezoelectric properties of synthetic hydroxyapatite-based organi-inorganic hydrated materials. Results in Physics. 2016;**6**:925-932

[43] More N, Kapusetti G. Piezoelectric material – A promising approach for bone and cartilage regeneration. Medical Hypotheses. 2017;**108**:10-16

[44] Kholkin A, Amdursky N, Bdikin I, Gazit E, Rosenman G. Strong piezoelectricity in bioinspired peptide nanotubes. ACS Nano. 2010;**4**(2):610-614

[45] Shin DM, Han HJ, Kim WG, Kim E, Kim C, Hong S, et al. Bioinspired piezoelectric nanogenerators based on vertically aligned phage nanopillars. Energy & Environmental Science. 2015;**8**:3198-3203

[46] Weitzel EK, Tasker R, Brownell WE. Outer hair cell piezoelectricity: Frequency response enhancement and resonance behavior. The Journal of the Acoustical Society of America. 2003;**114**(3):1462-1466

[47] Inaoka T, Shintaku H, Nakagawa T, Kawano S, Ogita H, Sakamoto T, et al. Piezoelectric materials mimic the function of the cochlear sensory epithelium. Proceedings of the National Academy of Sciences. 2011;**108**(45):18390-18395

[48] Liu XY, Fan F, Niu LL, Chen QN, Zheng HR, Li JY. Strong correlation between early stage atherosclerosis and electromechanical coupling of aorta. Nanoscale. 2016;**8**:6975-6980

[49] Nakamura M, Nagai A, Tanaka Y, Sekijima Y, Yamashita K. Polarized hydroxyapatite promotes spread and motility of osteoblastic cells. Journal of Biomedical Materials Research. Part A. 2010;**92**(2):783-790

[50] Rajabi AH, Jaffe M, Arinzeh TL. Piezoelectric materials for tissue regeneration: A review. Acta Biomaterialia. 2015;**24**:12-23

[51] Gautam CR, Kumar S, Biradar S, Jose J, Mishra VK. Synthesis and enhanced mechanical properties of MgO substituted hydroxyapatite: A bone substitute material. RSC Advances. 2016;**6**:67565-67574

[52] Martins P, Lopes AC, Lanceros-Mendez S. Electroactive phases of poly(vinylidene fluoride): Determination, processing and applications. Progress in Polymer Science. Apr 2014;**2014**(39):683-706

[53] Dargahi J. A piezoelectric tactile sensor with three sensing elements for robotic, endoscopic and prosthetic applications. Sensors and Actuators, A: Physical. 2000;**80**(1):23-30

[54] Zhang Y, Zhang F, Zhu D. Advances of flexible pressure sensors toward artificial intelligence and health care applications. Materials Horizons. Mar 2015;**2015**(2):133-254

[55] Seneviratne S, Hu Y, Nguyen T, Lan G, Khalifa S, Thilakarathna K, et al. A survey of wearable devices and challenges. IEEE Communications Surveys & Tutorials. 2017;**19**(4): 2573-2620

[56] Ghosh SK, Mandal D. Bio-assembled, piezoelectric prawn shell made self-powered wearable sensor for noninvasive physiological signal monitoring. Applied Physics Letters. 2017;**110**:123701

[57] Liu Z, Zhang S, Jin YM, Ouyang H, Zou Y, Wang XX, et al. Flexible piezoelectric nanogenerator in wearable self-powered active sensor for respiration and healthcare monitoring. Semiconductor Science and Technology. 2017;**32**:064004

[58] Kwak J, Kingon A, Kim SH. Lead-free ($Na_{0.5}$, $K_{0.5}$)NbO_3 thin films for the implantable piezoelectric medical sensor applications. Materials Letters. 2012;**82**:130-132

[59] McLaughlin J, McNeill M, Braun B, McCormack PD. Piezoelectric sensor determination of arterial pulse wave velocity. Physiological Measurement. 2003;**24**:693-702

[60] Bu N, Ueno N, Fukuda O. Monitoring of respiration and heartbeat during sleep using a flexible piezoelectric film sensor and empirical mode decomposition. In: 29th Annual International Conference of the IEEE Engineering in Medicine and Biology Society (EMBS). Lyon, France: IEEE; 2007

[61] Chiu Y, Lin W, Wang H, Huang SB, Wu M. Development of a piezoelectric polyvinylidene fluoride (PVDF) polymer-based sensor patch for simultaneous heartbeat and respiration monitoring. Sensors and Actuators, A. 2013;**189**:328-334

[62] Ghosh SK, Adhikary P, Jana S, Biswas A, Sencadas V, Gupta SD, et al. Electrospun gelatin nanofiber based self-powered bio-e-skin for health care monitoring. Nano Energy. Jun 2017;**36**:166-175

[63] Lu C, Wu S, Lu B, Zhang Y, Du Y, Feng X. Ultrathin flexible piezoelectric sensors for monitoring eye fatigue. Journal of Micromechanics and Microengineering. 2018;**28**: 025010

[64] Curry EJ, Ke K, Chorsi M,Worbel KS, Miller AN, Patel A, et al. Biodegradable Piezoelectric Force Sensor. Proceedings of the National Academy of Sciences. 2018:**115**(5):909-914

[65] Markus DT, Hayes MC, inventors. Piezoelectric Sensor for Vision Correction. US patent 20160030160A1. Apr 25, 2017

[66] Sazonov ES, Fontana JM. A sensor system for automatic detection of food intake through non-invasive monitoring of chewing. IEEE Sensors Journal. 2012;**12**(5):1340-1348

[67] Sonar HA, Paik J. Soft pneumatic actuator skin with piezoelectric sensors for vibrotactile feedback. Frontiers in Robotics and AI. 2016;**2**(38):1-11

[68] Wang X, Dong L, Zhang H, Yu R, Pan C, Wang ZL. Recent progress in electronic skin. Advanced Science. 2015;**2**(10):1500169

[69] Park J, Kim M, Lee Y, Lee H, Ko H. Fingertip skin-inspired microstructured ferroelectric skins discriminate static/dynamic pressure and temperature stimuli. Science Advances. Oct 2015;**1**(9):e1500661

[70] Hwang SK, Hwang HY. Development of a tactile sensing system using piezoelectric robot skin materials. Smart Materials and Structures. 2013;**22**:055004

[71] Yoon J, Kim S, Kim D, Kim ID, Hong S, No K. Fabrication of highly-ordered and well-aligned PbTiO3/TiN core-shell nanotube arrays. Small. 2015;**11**(31):3750-3754

[72] Paik H, Choi YY, Hong S, No K. Effect of Ag nanoparticle concentration on the electrical and ferroelectric properties of Ag/P(VDF-TrFE) composite films. Scientific Reports. 2015; **5**:13209

[73] Lucklum R, Henning B, Hauptmann P, Schierbaum KD, Vaihinger S, Gopel W. Quartz microbalance sensors for gas detection. Sensors and Actuators, A. 1991;**27**(1–3):705-710

[74] Vashist SK, Vashist P. Recent advances in quartz crystal microbalance-based sensors. Journal of Sensors. 2011;**2011**:5714045

[75] Marx KA. Quartz crystal microbalance: A useful tool for studying thin polymer films and complex biomolecular systems at the solution – surface interface. Biomacromolecules. 2003;**4**(5):1099-1120

[76] Kanazawa KK, Gordon JG. Frequency of a a quartz microbalance in contact with liquid. Analytical Chemistry. 1985;**57**(8):1770-1771

[77] Lasky SJ, Buttry DA. Sensors based on biomolecules immobilized on the piezoelectric quartz crystal microbalance: Detection of glucose using hexokinase. In: Cemical Sensors and Microinstrumentation. Washington DC: American Chemical Society; 1989. pp. 237-246

[78] Tang D, Li Q, Tang J, Su B, Chen G. An enzyme-free quartz crystal microbalance biosensor for sensitive glucose detection in biological fluids based on glucose/dextran displacement approach. Analytica Chimica Acta. 2011;**686**(1–2):144-149

[79] Lee SH, Park TH. Recent advances in the development of bioelectronic nose. Biotechnology and Bioprocess Engineering. 2010;**15**:22-29

[80] Ko HJ, Park TH. Piezoelectric olfactory biosensor: Ligand specificity and dose-dependence of an olfactory receptor expressed in a heterologous cell system. Biosensors & Bioelectronics. 2005 January;**20**(7):1327-1332

[81] İlik B, Koyuncuoğlu A, Uluşan H, Chamanian S, Işık D, Şardan-Sukas Ö, Külah H. In: Proceedings of Thin Film PZT Acoustic Sensor for Fully Implantable Cochlear Implants. 2017;**1**:366

[82] Song WJ, Jang J, Kim S, Choi H. Piezoelectric performance of continuous beam and narrow supported beam arrays for artificial basilar membranes. Electronic Materials Letters. 2014;**10**(5):1011-1018

[83] Jang J, Jang JH, Choi H. Biomimetic Artificial Basilar Membranes. Advanced Healthcare Materials. 2017;**6**(21):1700674

[84] Creighton FP, Guan X, Park S, Kymissis I, Nakajima HH, Olson ES. An Intracochlear pressure sensor as a microphone for a fully implantable cochlear implant. Otology & Neurotology. 2016;**37**(10):1596-1600

[85] Jung Y, Kim S, Kwak J, Kang H, Lee YH, Park S, et al. Development and characterization of piezoelectric artificial cochlear with micro actuator mimicking human cochlear. Journal of Physics Conference Series 2013;**476**:012015

[86] Deepa D, Jain G, Bansal T. Piezosurgery in dentistry. Journal of Oral Research and Review. 2016;**8**(1):27-31

[87] Pritchard J, Bowen CR, Lowrie F. Multilayer actuators: Review. British Ceramic Transactions. 2001;**100**(6):265-273

[88] Lucas M, Gachagan A, Cardon A. Research applications and opportunities in power ultrasonics. Proceedings of the IMechE. 2009;**233**:2949-2965

[89] Pereira CCS, Gealh WC, Meorin-Nogueira L, Garcia-Junior IR, Okamoto R. Piezosurgery applied to implant dentistry: Clinical and biological aspects. The Journal of Oral Implantology. 2014;**60**:401-408

[90] Labanca M, Azzola F, Vinici R, Rodella LF. Piezoelectric surgery: Twenty years of use. British Journal of Oral and Maxillofacial Surgery. 2008;**46**:265-269

[91] Leclercq P, Zenati C, Doban DM. Ultrasonic bone cut part 2: State-of-the-art specific clinical applications. Journal of Oral & Maxillofacial Surgery. 2008;**66**:183-188

[92] Degerliyurt K, Akar V, Denizci S, Yucel E. Bone lid technique with piezosurgery to preserve inferior alveolar nerve. OOOOE Journal. 2009;**108**(6):e1-e5

[93] Engelke D, Oehme B, Strackeljan J. A novel drive option for piezoelectric ultrasonic transducers. Modelling and Simulation in Engineering. 2011;**2011**:910876

[94] Arabaci T, Cicek Y, Canakci CF. Sonic and ultrasonic scalers in periodontal treatment: A review. International Journal of Dental Hygiene. 2007;**5**:2-12

[95] Vyas N, Pecheva E, Dehghani H, Sammons RL, Wang QX, Leppinen DM, et al. High speed imaging of Cavitation around dental ultrasonic Scaler tips. PLoS One. 2016;**11**(3): e0149804

[96] Felver B, King D, Lea S, Price G, Walmsley AD. Cavitation occurrence around ultrasonic dental scalers. Ultrasonics Sonochemistry. 2009;**16**(5):692-697

[97] Walmsley AD, Laird WRE, Lumley PJ. Ultrasound in dentistry part 2- periodontology and endodontics. Journal of Dentistry. 1992;**1992**:11-17

[98] Flemmig TF, Petersilka GJ, Mehl A, Hickel R, Klaiber B. Working parameters of a Magnetostrictive ultrasonic sealer influencing root substance removal in vitro. Journal of Periodontology. 1998;**69**(5):547-533

[99] Ianchulev T, Weinreb R, Tsai J, Lin S, Pasquale LR. High-precision piezo-ejection ocular microdosing: Phase II study on local and systemic effects of topical phenylephrine. Therapeutic Delivery. 2018;**9**(1):17-27

[100] Corcoran TK, Venkataramanan R, Hoffman RM, George MP, Petrov A, Richards T, et al. Systemic delivery of atropine sulfate by the microdose dry-powder inhaler. Journal of Aerosol Medicine and Pulmonary Drug Delivery. 2013;**26**(1):46-55

[101] Wang H, Zhang T, Zhao M. Micro-dosing of fine cohesive powders actuated by pulse inertia force. Micromachines. 2018;**9**(2):73

[102] Kar S, McWhorter S, Ford SM, Soper SA. Piezoelectric mechanical pump with nanoliter per minute pulse-free flow delivery for pressure pumping in micro-channels. The Analyst. 1998;**123**:1435-1441

[103] de Lima CR, Vatanabe SL, Choi A, Nakasone PH, Pires RF, Silva EC. A biomimetic piezoelectric pump: Computational and experimental characterization. Sensors and Actuators, A: Physical. 2009;**152**(1):110-118

[104] Liu G, Shen C, Yang Z, Cai X, Zhang H. A disposable piezoelectric micropump with high performance for closed-loop insulin therapy system. Sensors and Actuators, A: Physical. 2010;**163**(1):291-296

[105] Jang LS, Kan WH. Peristaltic piezoelectric micropump system for biomedical applications. Biomedical Microdevices. 2007;**9**(4):619-626

[106] Yildirim YA, Toprak A, Tigli O. Piezoelectric membrane actuators for micropump applications using PVDF-TrFE. Journal of Microelectromechanical Systems. 2018;**27**(1):86-94

[107] Mitcheson PD. Energy harvesting for human wearable and implantable bio-sensors. In: Annual International Conference of the IEEE EMBS; Buenos Aires, Argentina. 2010

[108] Karami MA, Inman D. Powering pacemakers from heartbeat vibrations using linear and nonlinear energy harvesters. Applied Physics Letters. 2011;**100**(4):042901

[109] Kim M, Hong S, Miller DJ, Dugundji J, Wardle BL. Size effect of flexible proof mass on the mechanical behavior of micron-scale cantilevers for energy harvesting applications. Applied Physics Letters. 2011;**99**:243106

[110] Hwang GT, Byun M, Jeong CK, Lee KJ. Flexible piezoelectric thin-film energy harvesters and nanosensors for biomedical applications. Advanced Healthcare Materials. 2015;**4**: 646-658

[111] Wahbah M, Alhawari M, Mohammad B, Saleh H, Ismail M. Characterization of human body-based thermal and vibration energy harvesting for wearable devices. IEEE Journal on Emerging and Selected Topics in Circuits and Systems. 2014;**4**(3):354-363

[112] Almouahed S, Gouriou M, Hamitouche C, Stindel E, Roux C. The use of piezoceramics as electrical energy harvesters within instrumented knee implant during walking. IEEE/ ASME Transactions on Mechatronics. 2011;**16**(5):799-807

[113] Wong YLL. Piezoelectric ribbons for stretchable energy harvesting [PhD dissertation]. Pittsburgh (PA): Carnegie Mellon University; 2016

[114] Safaei M, Meneghini RM, Anton SR. Energy harvesting and sensing with embedded piezoelectric ceramics in knee implants. IEEE/ASME Transactions on Mechatronics. 2018;**99**:1

[115] Shafer MW, Garcia E. The power and efficiency limits of piezoelectric energy harvesting. Journal of Vibration and Acoustics. Apr 2014;**136**(2):021007

[116] Hong S, Tong S, Park WI, Hiranaga Y, Cho Y, Roelofs A. Charge gradient microscopy. Proceedings of the National Academy of Sciences of the United States of America. 2014; **111**:6566-6569

[117] Tong S, Park WI, Choi YY, Stan L, Hong S, Roelofs A. Mechanical removal and rescreening of local screening charges at ferroelectric surfaces. Physical Review Applied. 2015;**3**:014003

[118] Tong S, Jung IW, Choi YY, Hong S, Roelofs A. Imaging ferroelectric domains and domain walls using charge gradient microscopy: Role of screening charges. ACS Nano. 2016;**10**: 2568-2574

[119] Hong S, Nakhmanson SM, Fong DD. Screening mechanisms at polar oxide heterointerfaces. Reports on Progress in Physics. 2016;**79**:076501